Wavelets for Computer Graphics
Theory and Applications

The Morgan Kaufmann Series in Computer Graphics and Geometric Modeling
Series Editor, Brian A. Barsky

Wavelets for Computer Graphics: Theory and Applications
    Eric J. Stollnitz, Tony D. DeRose, and David H. Salesin

Jim Blinn's Corner: A Trip Down the Graphics Pipeline
    Jim Blinn

Interactive Curves and Surfaces: A Multimedia Tutorial of CAGD
    Alyn Rockwood and Peter Chambers

Principles of Digital Image Synthesis
    Andrew S. Glassner

Radiosity & Global Illumination
    François X. Sillion and Claude Puech

Knotty: A B-Spline Visualization Program
    Jonathan Yen

User Interface Management Systems: Models and Algorithms
    Dan R. Olsen, Jr.

Making Them Move: Mechanics, Control, and Animation of Articulated Figures
    Edited by Norman I. Badler, Brian A. Barsky, and David Zeltzer

Geometric and Solid Modeling: An Introduction
    Christoph M. Hoffmann

An Introduction to Splines for Use in Computer Graphics and Geometric Modeling
    Richard H. Bartels, John C. Beatty, and Brian A. Barsky

Wavelets for Computer Graphics
Theory and Applications

Eric J. Stollnitz
University of Washington

Tony D. DeRose
Pixar Animation Studios

David H. Salesin
University of Washington

Morgan Kaufmann Publishers, Inc.
San Francisco, California

Sponsoring Editor Michael B. Morgan
Production Manager Yonie Overton
Production Editor Cheri Palmer
Editorial Coordinator Marilyn Alan
Cover and Color Insert Design Ross Carron
   Design
Cover photograph Bryan Whitney/
   PHOTONICA

Text Design and Composition Professional
   Book Center
Chapter and Part Opener Art Eric Stollnitz
   and Brad West
Copyeditor Erin Milnes
Proofreader Jennifer McClain
Indexer Steve Rath
Printer Courier Corporation

This book was composed in Corel Ventura Publisher in Myriad Headline, Formata, and Times Roman.

Morgan Kaufmann Publishers, Inc.
Editorial and Sales Office
340 Pine Street, Sixth Floor
San Francisco, CA 94104-3205
USA
Telephone 415 / 392-2665
Facsimile 415 / 982-2665
E-mail mkp@mkp.com
Web site http://www.mkp.com

© 1996 by Morgan Kaufmann Publishers, Inc.

Library of Congress Cataloging-in-Publication Data

Stollnitz, Eric J.
   Wavelets for computer graphics : theory and applications / Eric J.
Stollnitz, Tony D. DeRose, David H. Salesin.
      p. cm.
   Includes bibliographical references and index.
   ISBN 1-55860-375-1
   1. Computer graphics. 2. Wavelets (Mathematics) I. DeRose, Tony
D. II. Salesin, David H. III. Title.
T385.S796   1996
006.6'01'5152433—dc20   96-7950
                                                CIP

*We dedicate this book
to Janet and Fred Stollnitz,
to Cindy DeRose and Carolyn and Gene Anderson,
and to Andrea Lingenfelter and Rudy and Gene Salesin.*

FOREWORD

Ingrid Daubechies, Princeton University

In the 1980s, wavelets were born as an alternative to the windowed Fourier transform for signal analysis. It became apparent soon that they were much more—they were in fact a reincarnation of ideas that had existed earlier in many other fields. This is the fate of many a "new" idea. In the case of wavelets, the range of different fields to which they turned out to be connected is surprisingly broad, encompassing, among others, hard estimates in pure mathematics, the renormalization group concept in physics, subband filter banks in electrical engineering, and subdivision schemes in computer science.

More than just a synthesis of ideas from these widely different fields, wavelets added new facets to each of them, providing new insights as well as simplifying older approaches. Regardless of their background, many wavelet researchers found themselves enjoying the contact with other fields to which their study of wavelets unwittingly led them.

Coming to wavelets from a physics and mathematics background, I found great pleasure in reading in this book about the many uses of wavelets in computer graphics, an area with which I am less familiar. From applications of Haar wavelets in image editing or image querying, and of smoother wavelets for the construction of curves and surfaces, to the use of wavelets in physical simulation, where they come closer to mathematical analysis again, the course through wavelet applications charted in this book is multifaceted and always interesting.

Whether you are a novice to wavelets or not, I invite you to follow the authors on the trip they have planned—you will find it a rewarding experience, as I did.

CONTENTS

Color Plates   following page 102

Foreword by Ingrid Daubechies   vii

List of Figures   xv

Preface   xxi

Notation   xxv

1   INTRODUCTION   1

   1.1   Multiresolution methods   2

   1.2   Historical perspective   3

   1.3   Overview of the book   5

I   IMAGES

2   HAAR: THE SIMPLEST WAVELET BASIS   9

   2.1   The one-dimensional Haar wavelet transform   9

   2.2   One-dimensional Haar basis functions   11

   2.3   Orthogonality and normalization   16

   2.4   Wavelet compression   18

3   IMAGE COMPRESSION   21

3.1   Two-dimensional Haar wavelet transforms   21

3.2   Two-dimensional Haar basis functions   25

3.3   Wavelet image compression   28

3.4   Color images   30

3.5   Summary   31

4   IMAGE EDITING   33

4.1   Multiresolution image data structures   34

4.2   Image editing algorithm   36

4.3   Boundary conditions   39

4.4   Display and editing at fractional resolutions   40

4.5   Image editing examples   41

5   IMAGE QUERYING   43

5.1   Image querying by content   45

5.2   Developing a metric for image querying   46

5.3   Image querying algorithm   50

5.4   Image querying examples   53

5.5   Extensions   55

II   CURVES

6   SUBDIVISION CURVES   61

6.1   Uniform subdivision   62

6.2   Nonuniform subdivision   66

6.3   Evaluation masks   68

6.4   Nested spaces and refinable scaling functions   72

7   THE THEORY OF MULTIRESOLUTION ANALYSIS   79

7.1   Multiresolution analysis   80

7.2   Orthogonal wavelets   85

7.3   Semiorthogonal wavelets   89

7.4   Biorthogonal wavelets   97

7.5   Summary   107

8   MULTIRESOLUTION CURVES   109

8.1   Related curve representations   110

8.2   Smoothing a curve   111

8.3   Editing a curve   112

8.4   Scan conversion and curve compression   120

9   MULTIRESOLUTION TILING   125

9.1   Previous solutions to the tiling problem   126

9.2   The multiresolution tiling algorithm   129

9.3   Time complexity   135

9.4   Tiling examples   136

III   SURFACES

10   SURFACE WAVELETS   141

10.1 Overview of multiresolution analysis for surfaces   142

10.2 Subdivision surfaces   143

10.3 Selecting an inner product   151

10.4 A biorthogonal surface wavelet construction   152

10.5 Multiresolution representations of surfaces   158

11   SURFACE APPLICATIONS   161

11.1 Conversion to multiresolution form   161

11.2 Surface compression   163

11.3 Continuous level-of-detail control   164

11.4 Progressive transmission   165

11.5 Multiresolution editing   165

11.6 Future directions for surface wavelets   166

IV   PHYSICAL SIMULATION

12   VARIATIONAL MODELING   171

12.1 Setting up the objective function   172

12.2 The finite-element method   173

12.3 Using finite elements in variational modeling   173

12.4 Variational modeling using wavelets   177

12.5 Adaptive variational modeling   179

13   GLOBAL ILLUMINATION   181

13.1 Radiosity   181

13.2 Finite elements and radiosity   183

13.3 Wavelet radiosity   186

13.4 Enhancements to wavelet radiosity   192

14   FURTHER READING   195

14.1 Theory of multiresolution analysis   195

14.2 Image applications   197

14.3 Curve and surface applications   198

14.4 Physical simulation   199

V   APPENDICES

A   LINEAR ALGEBRA REVIEW   203

A.1 Vector spaces   203

A.2 Bases and dimension   204

A.3 Inner products and orthogonality   205

A.4 Norms and normalization   206

A.5 Eigenvectors and eigenvalues   207

B   B-SPLINE WAVELET MATRICES   209

B.1 Haar wavelets   210

B.2 Endpoint-interpolating linear B-spline wavelets   211

B.3 Endpoint-interpolating quadratic B-spline wavelets   212

B.4 Endpoint-interpolating cubic B-spline wavelets   214

C   MATLAB CODE FOR B-SPLINE WAVELETS   217

Bibliography 223

Index 235

LIST OF FIGURES

1.1   Selected applications of wavelets   **2**
     (a) typical time-series data
     (b) a cross-sectional contour through the human cerebral cortex
          Meyers (1994), fig. 2, p. 326. © 1994 Blackwell Publishers, Cambridge, MA.
     (c) a typical image
          "La promenade" by Pierre-August Renoir
     (d) a typical surface
          Eck et al. (1995), Color Plate 1k, p. 181. © 1995 Association for Computing Machinery, Inc.
1.2   Shift-invariant and shift-variant multiresolution analysis   **4**
2.1   Function approximation   **12**
2.2   The box basis for V^2 **13**
          Eric J. Stollnitz, Tony D. DeRose, and David H. Salesin. Wavelets for computer graphics: A primer, Part 1. IEEE Computer Graphics and Applications 15(3):77, fig. 2, May 1995. IEEE Computer Society Press, Los Alamitos, CA © 1995 Institute of Electrical and Electronics Engineers, Inc.
2.3   The Haar wavelets for W^1 **15**
          Eric J. Stollnitz, Tony D. DeRose, and David H. Salesin. Wavelets for computer graphics: A primer, Part 1. IEEE Computer Graphics and Applications 15(3):78, fig. 3, May 1995. IEEE Computer Society Press, Los Alamitos, CA. © 1995 Institute of Electrical and Electronics Engineers, Inc.
2.4   Coarse approximations using L^2 compression   **20**
          Eric J. Stollnitz, Tony D. DeRose, and David H. Salesin. Wavelets for computer graphics: A primer, Part 1. IEEE Computer Graphics and Applications 15(3):81, fig. 4, May 1995. IEEE Computer Society Press, Los Alamitos, CA. © 1995 Institute of Electrical and Electronics Engineers, Inc.

3.1   Standard decomposition of an image   **22**

> *Eric J. Stollnitz, Tony D. DeRose, and David H. Salesin. Wavelets for computer graphics: A primer, Part 1.* IEEE Computer Graphics and Applications *15(3):82, fig. 5, May 1995. IEEE Computer Society Press, Los Alamitos, CA. © 1995 Institute of Electrical and Electronics Engineers, Inc.; "Mona Lisa," by Leonardo da Vinci.*

3.2   Nonstandard decomposition of an image   **24**

> *Eric J. Stollnitz, Tony D. DeRose, and David H. Salesin. Wavelets for computer graphics: A primer, Part 1.* IEEE Computer Graphics and Applications *15(3):82, fig. 6, May 1995. IEEE Computer Society Press, Los Alamitos, CA. © 1995 Institute of Electrical and Electronics Engineers, Inc.*

3.3   The standard construction of a two-dimensional Haar wavelet basis for V^2 **26**

> *Eric J. Stollnitz, Tony D. DeRose, and David H. Salesin. Wavelets for computer graphics: A primer, Part 1.* IEEE Computer Graphics and Applications *15(3):83, fig. 7, May 1995. IEEE Computer Society Press, Los Alamitos, CA. © 1995 Institute of Electrical and Electronics Engineers, Inc.*

3.4   The nonstandard construction of a two-dimensional Haar wavelet basis for V^2 **27**

> *Eric J. Stollnitz, Tony D. DeRose, and David H. Salesin. Wavelets for computer graphics: A primer, Part 1.* IEEE Computer Graphics and Applications *15(3):83, fig. 8, May 1995. IEEE Computer Society Press, Los Alamitos, CA. © 1995 Institute of Electrical and Electronics Engineers, Inc.*

3.5 L^2 Haar wavelet image compression   **30**

> *Eric J. Stollnitz, Tony D. DeRose, and David H. Salesin. Wavelets for computer graphics: A primer, Part 1.* IEEE Computer Graphics and Applications *15(3):84, fig. 9, May 1995. IEEE Computer Society Press, Los Alamitos, CA. © 1995 Institute of Electrical and Electronics Engineers, Inc.*

4.1   Nonstandard Haar wavelets   **35**
5.1   Preprocessing steps   **51**

> (a) *"Vase of Irises," Vincent van Gogh*
> (b–d) *Courtesy of Adam Finkelstein, University of Washington*

5.2   Comparing the signature of a query to the signature of a target image   **52**

> *Courtesy of Adam Finkelstein, University of Washington*

5.3   Query accuracy graph   **55**
5.4   Query speed graph   **56**
6.1   Chaikin's algorithm for a function   **63**
6.2   Splitting step   **64**
6.3   Chaikin's algorithm for a closed parametric curve   **65**
6.4   The Daubechies subdivision scheme   **66**
6.5   The DLG interpolating scheme for a closed parametric curve   **67**
6.6   The tracking of a vertex c^0 of f^0 through the subdivision process   **70**
6.7   Box function refinement   **74**
6.8   Endpoint-interpolating cubic B-spline scaling functions generated by repeated subdivision   **75**
6.9   Uniform subdivision produces scaling functions that are shifted copies of one another   **76**
7.1   The filter bank   **84**

> *Eric J. Stollnitz, Tony D. DeRose, and David H. Salesin. Wavelets for computer graphics: A primer, Part 2.* IEEE Computer Graphics and Applications *15(4):77, fig. 1, July 1994.*

IEEE Computer Society Press, Los Alamitos, CA. © 1995 Institute of Electrical and Electronics Engineers, Inc.

7.2 Daubechies basis functions **89**

7.3 Endpoint-interpolating B-spline scaling functions **92**
Eric J. Stollnitz, Tony D. DeRose, and David H. Salesin. Wavelets for computer graphics: A primer, Part 2. IEEE Computer Graphics and Applications 15(4):78, fig. 2, July 1994. IEEE Computer Society Press, Los Alamitos, CA. © 1995 Institute of Electrical and Electronics Engineers, Inc.

7.4 Endpoint-interpolating B-spline wavelets **96**

7.5 Linear lazy wavelet **103**

7.6 Single-knot wavelet **105**

8.1 Smoothing a curve continuously **112**
Finkelstein and Salesin (1994), fig. 2, p. 263. © 1994 Association for Computing Machinery, Inc.

8.2 Changing the overall sweep of a curve without affecting its character **113**
Finkelstein and Salesin (1994), fig. 3, p. 264. © 1994 Association for Computing Machinery, Inc.

8.3 The middle of the dark curve is pulled **114**
Finkelstein and Salesin (1994), fig. 4, p. 264. © 1994 Association for Computing Machinery, Inc.

8.4 A curve with parameterization that changes most rapidly in the middle **118**
Finkelstein and Salesin (1994), fig. 5, p. 265. © 1994 Association for Computing Machinery, Inc.

8.5 Changing the character of a curve without affecting its sweep **119**
Finkelstein and Salesin (1994), fig. 6, p. 265. © 1994 Association for Computing Machinery, Inc.

8.6 Orientation of edits **120**
Finkelstein and Salesin (1994), fig. 7, p. 265. © 1994 Association for Computing Machinery, Inc.

8.7 Scan-converting a curve within a guaranteed maximum error tolerance **123**
Finkelstein and Salesin (1994), figs. 8 and 9, p. 266. © 1994 Association for Computing Machinery, Inc.

9.1 Brain contours/tiling **126**
Meyers (1994), fig. 1, p. 326. © 1994 Blackwell Publishers, Cambridge, MA.

9.2 Tiling and graph problem **127**

9.3 (a) Brain contours **128**
Meyers (1994), fig. 2, p. 326. © 1994 Blackwell Publishers, Cambridge, MA.
(b) Optimized tiling **128**
Meyers (1994), fig. 3, p. 327. © 1994 Blackwell Publishers, Cambridge, MA.

9.4 Tilings of the contours in Figure 9.3 produced by linear-time methods **129**
Meyers (1994), fig. 11, p. 335. © 1994 Blackwell Publishers, Cambridge, MA.

9.5 The main steps of the multiresolution tiling algorithm **131**
Meyers (1994), fig. 1, p. 326. © 1994 Blackwell Publishers, Cambridge, MA.

9.6 Filter-bank reconstruction **132**

9.7 Illustration of a single-wavelet reconstruction in one dimension **134**

9.8   Edge swapping   **135**

9.9   A tiling of the contours in Figure 9.3   **136**
          Meyers (1994), fig. 12, p. 336; fig. 3, p. 327. © 1994 Blackwell Publishers, Cambridge,
          MA.

9.10   Tilings of the contours in Figure 9.3 computed by the single-wavelet multiresolution algorithm
          with a compression threshold   **137**
          Meyers (1994), fig. 12, p. 336. © 1994 Blackwell Publishers, Cambridge, MA.

10.1   Decomposition of a polyhedral surface   **143**
          Michael Lounsbery, Tony DeRose, and Joe Warren. Multiresolution surfaces of arbitrary
          topological type. ACM Transactions on Graphics *(to appear). ACM, New York. © 1996 As-*
          sociation for Computing Machinery, Inc.

10.2   Catmull-Clark subdivision   **144**

10.3   Doo-Sabin subdivision   **145**

10.4   The splitting step for triangular face schemes   **146**

10.5   Loop's subdivision scheme   **147**

10.6   Masks associated with Loop's scheme   **147**

10.7   The average mask used to compute an edge midpoint in the butterfly scheme   **149**

10.8   Surface tracking   **150**
          Michael Lounsbery, Tony DeRose, and Joe Warren. Multiresolution surfaces of arbitrary
          topological type. ACM Transactions on Graphics *(to appear). ACM, New York. © 1996 As-*
          sociation for Computing Machinery, Inc.

10.9   Polyhedral subdivision of a tetrahedron   **153**
          Michael Lounsbery, Tony DeRose, and Joe Warren. Multiresolution surfaces of arbitrary
          topological type. ACM Transactions on Graphics, *1996 (to appear). ACM, New York. ©*
          1996 Association for Computing Machinery, Inc.

10.10 Vertices within the k-discs of two edge endpoints in a regular triangulation   **156**

10.11 Polyhedral surface wavelets   **157**

11.1   (a) Complex mesh   **162**
          Eck et al. (1995), Color Plate 1k, p. 181. © 1995 Association for Computing Machinery,
          Inc.
          (b) Base mesh
          Eck et al. (1995), Color Plate 1j, p. 181. © 1995 Association for Computing Machinery,
          Inc.
          (c) Projection of the model into $V^5(M^0)$
          Eck et al. (1995), Color Plate 2b, p. 182. © 1995 Association for Computing Machinery,
          Inc.

11.2   Progressive transmission   **166**
          (a) *Eck et al. (1995), Color Plate 2h, p. 182. © 1995 Association for Computing Machin-*
          ery, Inc.
          (b) *Eck et al. (1995), Color Plate 2i, p. 182. © 1995 Association for Computing Machinery,*
          Inc.
          (c) *Courtesy of Matthias Eck, Technische Hochschule Darmstadt, Germany*

12.1   Quartic functions satisfying three interpolatory constraints   **175**

12.2 A sequence of iterations converging to the minimum-energy curve satisfying three constraints **178**

Gortler and Cohen (1995), fig. 1, p. 36. © 1993 Association for Computing Machinery, Inc. Reprinted with permission.

13.1 The geometry involved in radiosity transport from point x to point y **183**

13.2 Links to be considered when refining an existing link between basis functions **190**

Per H. Christensen, Eric J. Stollnitz, David H. Salesin, and Tony D. DeRose. Global illumination of glossy environments using wavelets and importance. ACM Transactions on Graphics, January 1996, fig. 11. ACM, New York. © 1996 Association for Computing Machinery, Inc.

B.1 The piecewise-constant B-spline scaling functions and wavelets for $j = 3$ **210**

Eric J. Stollnitz, Tony D. DeRose, and David H. Salesin. Wavelets for computer graphics: A primer, Part 2. IEEE Computer Graphics and Applications 15(4):82, fig. A, July 1994. IEEE Computer Society Press, Los Alamitos, CA. © 1995 Institute of Electrical and Electronics Engineers, Inc.

B.2 The linear B-spline scaling functions and wavelets for $j = 3$ **211**

Eric J. Stollnitz, Tony D. DeRose, and David H. Salesin. Wavelets for computer graphics: A primer, Part 2. IEEE Computer Graphics and Applications 15(4):82, fig. B, July 1994. IEEE Computer Society Press, Los Alamitos, CA. © 1995 Institute of Electrical and Electronics Engineers, Inc.

B.3 The quadratic B-spline scaling functions and wavelets for $j = 3$ **213**

Eric J. Stollnitz, Tony D. DeRose, and David H. Salesin. Wavelets for computer graphics: A primer, Part 2. IEEE Computer Graphics and Applications 15(4):83, fig. C, July 1994. IEEE Computer Society Press, Los Alamitos, CA. © 1995 Institute of Electrical and Electronics Engineers, Inc.

B.4 The cubic B-spline scaling functions and wavelets for $j = 3$ **215**

Eric J. Stollnitz, Tony D. DeRose, and David H. Salesin. Wavelets for computer graphics: A primer, Part 2. IEEE Computer Graphics and Applications 15(4):84, fig. D, July 1994. IEEE Computer Society Press, Los Alamitos, CA. © 1995 Institute of Electrical and Electronics Engineers, Inc.

COLOR PLATES

1 Compressing a color image (from Renoir's painting "La promenade") using the nonstandard Haar basis and the L^2 norm

2 Adding makeup and a glint in the eye to da Vinci's "Mona Lisa"

Berman et al. (1994), figs. 1a–d, p. 90. © 1994 Association of Computing Machinery, Inc.

3 Painting "in" and "under" at low resolutions

Berman et al. (1994), figs. 1e–h, p. 90. © 1994 Association of Computing Machinery, Inc.

4 Adding smog to an image with a range in scale of 100,000 to 1

Berman et al. (1994), figs. 1i–l, p. 90. © 1994 Association of Computing Machinery, Inc.

5 The image querying application

Jacobs et al. (1995), fig. 1, p. 281. © 1995 Association of Computing Machinery, Inc.

6 Queries and their target

Jacobs et al. (1995), fig. 2, p. 281. © 1995 Association of Computing Machinery, Inc. "Sunflowers" by Vincent van Gogh.

7   Progression of an interactive query
         Jacobs et al. (1995), fig. 3, p. 282. © 1995 Association of Computing Machinery, Inc.

8   A tensor-product spline surface, manipulated at different levels of detail
         Courtesy of Sean Anderson, Stanford University. (Black and white version first appeared in Finkelstein and Salesin [1994], fig. 11, p. 267. © 1994 Association of Computing Machinery, Inc.)

9   Wavelet approximations of a polyhedron
         Michael Lounsbery, Tony DeRose, and Joe Warren. Multiresolution surfaces of arbitrary topological type. ACM Transactions on Graphics, *1996 (to appear). ACM, New York. © 1996 Association for Computing Machinery, Inc.*

10   Compressed wavelet approximations of a bunny model
         Eck et al. (1995), Color Plates 2a–c, p. 182. © 1995 Association of Computing Machinery, Inc.

11   Compression and multiresolution editing of a smooth surface
         Michael Lounsbery, Tony DeRose, and Joe Warren. Multiresolution surfaces of arbitrary topological type. ACM Transactions on Graphics, *1996 (to appear). ACM, New York. © 1996 Association for Computing Machinery, Inc.*

12   Approximating color as a function over the sphere
         Michael Lounsbery, Tony DeRose, and Joe Warren. Multiresolution surfaces of arbitrary topological type. ACM Transactions on Graphics, *1996 (to appear). ACM, New York. © 1996 Association for Computing Machinery, Inc.*

13   Examples of surfaces minimizing an energy functional while meeting user-imposed constraints
         Gortler and Cohen (1995), Color Plate, p. 205. © 1993 Association for Computing Machinery, Inc. Reprinted with permission.

14   Radiosity solutions computed by a wavelet radiosity algorithm
         Courtesy of Dani Lischinski, University of Washington

15   Radiance solutions computed by a wavelet radiance algorithm
         Per H. Christensen, Dani Lischinski, Eric J. Stollnitz, and David H. Salesin. Clustering for glossy global illumination. ACM Transactions on Graphics, *fig. 9, 1996 (to appear). ACM, New York.*

PREFACE

Wavelets are a mathematical tool for hierarchically decomposing functions. Though rooted in approximation theory, signal processing, and physics, wavelets have also recently been applied to many problems in computer graphics. These graphics applications include image editing and compression, automatic level-of-detail control for editing and rendering curves and surfaces, surface reconstruction from contours, and fast methods for solving simulation problems in 3D modeling, global illumination, and animation.

Despite the growing evidence that wavelets are quickly becoming a core technique in computer graphics, most of the existing literature has been written primarily for the signal processing and approximation theory communities and is relatively inaccessible to researchers working in computer graphics. In addition, most of the established theory on wavelets has been developed for the theoretically pure case of signals of infinite length. Unfortunately, this classical theory begins to break down when it comes to representing the kinds of finite data sets—such as images, open curves, and bounded surfaces—that arise most commonly in computer graphics.

This book is intended to address both of these problems. First, it will provide the computer graphics professional and researcher with a firm understanding of the theory and applications of wavelets. The reader is assumed to have had a first course in linear algebra—but to have forgotten most of it (the text includes a linear algebra refresher with most of the necessary background in Appendix A). We have intentionally kept the style relatively light in tone, stressing intuition and clarity over rigor.

Second, the book takes an approach significantly different from that of existing texts, in that it focuses on a more generalized theory of wavelets, which allows wavelets to be constructed naturally on the kinds of bounded domains that arise quite commonly in computer graphics applications. This more generalized theory, it turns out, is intimately tied to the process of recursive subdivision. This text, therefore, develops the theory of subdivision curves and surfaces in its treatment of wavelets.

This book is by no means exhaustive, neither in its development of the theory of wavelets, nor in its coverage of their applications in computer graphics. Our goal in writing this text, rather, has been to emphasize the aspects of the theory that have already proven themselves to be most useful in computer graphics and to provide a small but sufficiently broad set of applications to illustrate how the theory can be applied in practice to a surprisingly wide variety of problems.

How this book came about

Our interest in wavelets began in late 1992 and early 1993. At that time wavelets had not yet (to our knowledge) been applied to problems in computer graphics, but we felt that there might be many such applications. The easiest way to learn a new area is to hold a graduate seminar on the topic, so that's what we did in the spring of 1993 with the help of Joe Warren. Within two or three weeks it became apparent that wavelets would have widespread use in computer graphics and geometric modeling. By the end of the quarter four students were off and running on Ph.D. topics (Per Christensen, Adam Finkelstein, Michael Lounsbery, and David Meyers), and one on a Master's project (Debbie Berman). Much of this work was later published in the proceedings of the annual SIGGRAPH and Graphics Interface conferences, as well as in recent issues of *IEEE Computer Graphics and Applications* and *ACM Transactions on Graphics.*

This text is also a direct outgrowth of that seminar. Although it draws heavily on our previously published work (including our tutorial article on wavelets and graphics [120]), we have added a considerable amount of new material, and we have attempted to develop a unified framework and consistent notation for all of the results.

Organization of the book

The book is organized into four main parts according to application area: images, curves, surfaces, and physical simulation. Within these parts, chapters on theory are intermingled with chapters on specific applications. We felt that such an organization would provide a more interesting narrative—and would also better motivate the theory—than segregating all the theory chapters from the practice.

While we would not recommend reading the applications chapters without the chapters on theory, it should be possible to do the reverse. Thus, readers who are interested in learning only about the theory of wavelets may wish to focus their attention on Chapters 2, 3, 6, 7, and 10.

Acknowledgments

We are grateful to Sean Anderson, Jason Bartell, Debbie Berman, Per Christensen, Michael Cohen, Tom Duchamp, Matthias Eck, Adam Finkelstein, Steven Gortler, Hugues Hoppe, Chuck Jacobs, Dani Lischinski, Michael Lounsbery, David Meyers, Werner Stuetzle, and Joe Warren for contributing material to the book. We are also indebted to the other members of the graphics research group at the University of Washington's Department of Computer Science and Engineering who helped make this book possible.

Thanks also to Ronen Barzel, Michael Shantzis, and Andy van Dam for reviewing preliminary chapters, and to Peter Shirley, François Sillion, and Richard Szeliski for reviewing early versions of the entire monograph and providing many helpful comments.

NOTATION

Although we are undoubtedly guilty of some lapses, we've attempted to use consistent notation throughout the book. By convention, we use uppercase boldface italics to denote matrices, lowercase boldface italics for vectors that are column matrices of scalars, and normal italics for other mathematical symbols. The symbols that appear most frequently are listed below, along with brief explanations of their meanings.

Symbol	Meaning
γ	parametric curve
σ	parametric surface
ϕ	scaling function
ψ	wavelet
$\boldsymbol{\Phi}^j$	row matrix of scaling functions
$\boldsymbol{\Psi}^j$	row matrix of wavelets
$\boldsymbol{A}^j$	scaling function analysis matrix
$\boldsymbol{B}^j$	wavelet analysis matrix
$\boldsymbol{c}^j$	scaling function coefficients
$\boldsymbol{d}^j$	wavelet coefficients
j	hierarchy level
$\boldsymbol{P}^j$	scaling function refinement and synthesis matrix
$\boldsymbol{Q}^j$	wavelet refinement and synthesis matrix

S	lifting matrix
u	general basis function
$\boldsymbol{u}$	row matrix of basis functions
V^j	scaling function space
W^j	wavelet space
$v(j)$	dimension of V^j
$w(j)$	dimension of W^j
$\langle f \mid g \rangle$	inner product of f and g
$[\langle \boldsymbol{f} \mid \boldsymbol{g} \rangle]$	the matrix whose (k, l) entry is $\langle f_k \mid g_l \rangle$

INTRODUCTION

1. Multiresolution methods — 2. Historical perspective — 3. Overview of the book

Central to virtually every scientific and engineering discipline is the need to analyze, visualize, and manipulate large quantities of data. The particular data in these applications may take many forms, as illustrated in Figure 1.1. They may be functions of a single parameter, such as the time-series data used in signal processing (a), or complex cross-sectional contours encountered in medical applications (b). The data may also be functions of two or more parameters, such as two-dimensional images (c), surface models (d), or higher-dimensional "global illumination" solutions for photorealistic lighting such as the one shown in Color Plate 15.

In all of these cases, the simplest way to represent the information is with a sequence of points. For example, a time series is most easily represented as a sequence (t_i, y_i). Each such point provides complete information about the behavior of the series at t_i but absolutely no information about the behavior of the series elsewhere. In contrast, many applications require an analysis of the series at broader scales. For example, a region of rapid change in the series can only be detected by examining many points at once.

The classic tool for addressing these issues is Fourier analysis, which can be used to convert point data into a form that is useful for analyzing frequencies. A difficulty with Fourier techniques, however, is that each Fourier coefficient contains complete information about the behavior of the series at one frequency but no information about its behavior at other

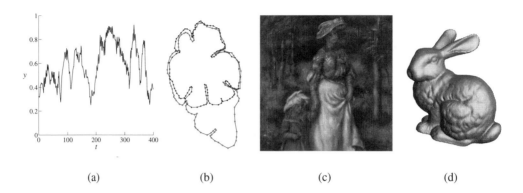

FIGURE 1.1   Selected applications of wavelets: (a) typical time-series data; (b) a cross-sectional contour through the human cerebral cortex; (c) a typical image; (d) a typical surface.

frequencies. Fourier techniques are also difficult to adapt to many situations of practical importance. For instance, most of the time series encountered in practice are finite and aperiodic, but the discrete Fourier transform can be applied only to periodic functions.

1.1   Multiresolution methods

The deficiencies of Fourier techniques have led researchers in a variety of disciplines (including approximation theory, physics, signal and image processing, as well as computer graphics) to develop various *hierarchical representations* of functions. The basic idea behind all hierarchical methods (also called *multiresolution methods*) is to represent functions with a collection of coefficients, each of which provides some limited information about both the position and the frequency of the function.

Although there are a wide variety of methods for representing functions in a hierarchical fashion, the recently developed theory of *wavelets* provides an extremely useful mathematical toolkit for hierarchically decomposing functions in ways that are both efficient and theoretically sound. Broadly speaking, a wavelet representation of a function consists of a coarse overall approximation together with detail coefficients that influence the function at various scales.

A central goal of this book is to demonstrate how wavelet representations are beginning to profoundly affect all areas of computer graphics, due in large part to the many useful properties associated with them. In addition to the hierarchical nature of wavelets, these properties include

- *Linear-time complexity.* Transforming to and from a wavelet representation can generally be accomplished in linear time, allowing for very fast algorithms.

- *Sparsity.* For functions typically encountered in practice, many of the coefficients in a wavelet representation are either zero or negligibly small. This property offers the opportunity both to compress data and to accelerate the convergence of iterative solution techniques.

- *Adaptability.* Unlike Fourier techniques, wavelets are remarkably flexible in that they can be adapted to represent a wide variety of functions, including functions with discontinuities, functions defined on bounded domains, and functions defined on domains of arbitrary topological type. Consequently, wavelets are equally well suited to problems involving images, open or closed curves, and surfaces of just about any variety.

While this book explores a number of applications of wavelets in detail, it is by no means intended to be comprehensive. Rather, the applications here are intended primarily to provide a broad set of examples of the many ways in which wavelets can be used. Indeed, the applications in this book cover areas as diverse as interactive techniques (Chapters 4, 5, 8, 9, 11, and 12), level-of-detail modeling and rendering (Chapters 8 and 11), combinatorial optimization (Chapter 9), continuous optimization (Chapter 12), and solving integral equations (Chapter 13).

1.2   Historical perspective

Wavelets have recently become enormously popular. However, their roots go back at least a century to the work of Karl Weierstrass [128], who in 1873 described a family of functions that are constructed by superimposing scaled copies of a given base function. The functions he defined are fractal, being everywhere continuous but nowhere differentiable. Another important early milestone was Alfred Haar's 1909 construction of the first orthonormal system of compactly supported functions, now called the *Haar basis* [53]. The Haar basis still serves as the foundation of modern wavelet theory, and we will study it in detail in the next chapter. Yet another significant advance came in 1946, when Dennis Gabor [44] described a nonorthogonal basis of (what are now called) wavelets with unbounded support, based on translated Gaussians.

The term *wavelet* comes from the field of seismology, where it was coined by Ricker in 1940 to describe the disturbance that proceeds outward from a sharp seismic impulse or explosive charge [102]. In 1982, Morlet et al. [86] showed how these seismic wavelets could be effectively modeled with the mathematical functions that Gabor had defined. Later, Grossman

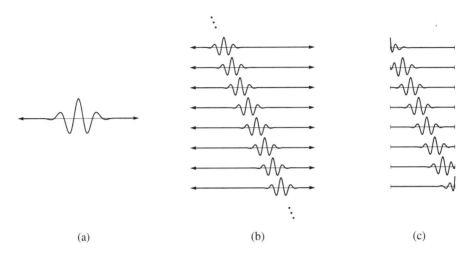

(a)   (b)   (c)

FIGURE 1.2   Shift-invariant and shift-variant multiresolution analysis: (a) a mother wavelet; (b) shift-invariant wavelets; (c) shift-variant wavelets.

and Morlet [52] showed how arbitrary signals could be analyzed in terms of scales and translates of a single *mother wavelet* function. Yves Meyer [80, 82] and Stephane Mallat [77] developed this notion into a theory called *multiresolution analysis.* In 1989, Mallat [78] showed how multiresolution analysis could be viewed as just another form of the *pyramid algorithms* used in image processing [6] and the *quadrature mirror filters* used in signal analysis [23, 116].

The now-classical form of multiresolution analysis, as put forth by Meyer and Mallat, decomposes signals onto a set of basis functions, called wavelets, in which every wavelet is just a scaled and translated copy of a single unique function, called the mother wavelet, as illustrated in Figure 1.2(a) and (b). We'll call this traditional approach a *shift-invariant* theory, since the wavelets lying on different parts of the unbounded real line all look the same. Unfortunately, the shift-invariant approach, while being remarkably beautiful from a theoretical standpoint, is problematic for most computer graphics applications since many functions of interest, such as images or open curves, are defined only on some *bounded* portion of the reals.

In this book, we present a somewhat different, more generalized version of multiresolution analysis than is found in other texts. This generalized version accommodates more naturally the kinds of finite data sets encountered in practical computer graphics applications. Unlike classical multiresolution analysis, the generalized theory we present here is *shift variant:* it accommodates bounded data sets by introducing different, specially adapted wavelets near the boundaries, as illustrated in Figure 1.2(c).

It turns out that the shift-variant multiresolution analysis described in this book is very closely related to the theory of recursive subdivision. Indeed, as we show in Chapters 6 and 7, the functions to which shift-variant multiresolution analysis can be applied turn out to be *exactly* those functions that can be generated through a subdivision process. Thus, this book also takes a very different approach in developing much of the theory behind wavelets. Unlike other texts, in which Fourier techniques are emphasized, in our development, the study, analysis, and construction of recursive subdivision schemes will play a central role.

1.3 Overview of the book

We begin Part I by examining how piecewise-constant functions and images can be represented hierarchically using the Haar basis (Chapter 2). This analysis leads us to the applications of image compression (Chapter 3), multiresolution image editing (Chapter 4), and multiresolution image querying (Chapter 5).

In Parts II and III, we tackle the subject of curves and surfaces.

We begin Part II with a discussion of *subdivision curves*—curves that are generated through the process of recursive subdivision (Chapter 6). We then present the formal framework of shift-variant multiresolution analysis (Chapter 7), which is intimately tied to subdivision. The theory presented in this chapter is then applied to the problems of multiresolution curve editing (Chapter 8) and multiresolution surface tiling from contours (Chapter 9).

Part III discusses how the framework can be generalized to surfaces of arbitrary topological type (Chapter 10) and discusses a number of applications, including surface compression, progressive transmission, and editing (Chapter 11).

Finally, Part IV focuses on wavelet-based algorithms for solving two physical simulation problems: variational modeling (Chapter 12) and global illumination (Chapter 13). The main body of the text closes with some directions for further reading (Chapter 14).

We have also included three appendices in Part V: a refresher on linear algebra (Appendix A); all of the matrices required to implement endpoint-interpolating B-spline wavelets of low degree (Appendix B); and Matlab code for generating these matrices (Appendix C).

IMAGES

HAAR: THE SIMPLEST WAVELET BASIS

1. The one-dimensional Haar wavelet transform — 2. One-dimensional Haar basis functions — 3. Orthogonality and normalization — 4. Wavelet compression

The Haar basis is the simplest wavelet basis. In this chapter, we will begin by examining how a one-dimensional function can be decomposed using Haar wavelets. We will then look at the Haar basis functions in detail and see how Haar wavelet decomposition can be used for compression. Later, in the following three chapters, we'll explore some applications of the Haar basis: image compression, image editing, and image querying.

2.1 The one-dimensional Haar wavelet transform

To get a sense for how wavelets work, let's start out with a simple example. Suppose we are given a one-dimensional "image" with a resolution of 4 pixels, having the following pixel values:

$$[9\ 7\ 3\ 5]$$

This image can be represented in the Haar basis, the simplest wavelet basis, by computing a wavelet transform as follows. Start by averaging the pixels together, pairwise, to get a new lower-resolution image with these pixel values:

$$[8\ 4]$$

Clearly, some information has been lost in this averaging and down-sampling process. In order to be able to recover the original four pixel values from the two averaged pixels, we need to store some *detail coefficients*, which capture that missing information. In our example, we will choose 1 for the first detail coefficient, since the average we computed is 1 less than 9 and 1 more than 7. This single number allows us to recover the first two pixels of our original four-pixel image. Similarly, the second detail coefficient is -1, since $4 + (-1) = 3$ and $4 - (-1) = 5$.

Summarizing, we have so far decomposed the original image into a lower-resolution (two-pixel) version and detail coefficients as follows:

Resolution	Averages	Detail Coefficients
4	[9 7 3 5]	
2	[8 4]	[1 −1]

Repeating this process recursively on the averages gives the full decomposition:

Resolution	Averages	Detail Coefficients
4	[9 7 3 5]	
2	[8 4]	[1 −1]
1	[6]	[2]

Finally, we will define the *wavelet transform* (also called the *wavelet decomposition*) of the original four-pixel image to be the single coefficient representing the overall average of the original image, followed by the detail coefficients in order of increasing resolution. Thus, for the one-dimensional Haar basis, the wavelet transform of our original four-pixel image is given by

$$[6\ 2\ 1\ -1]$$

The way we computed the wavelet transform, by recursively averaging and differencing coefficients, is called a *filter bank*—a process we will generalize to other types of wavelets in Chapter 7. Note that no information has been gained or lost by this process: The original image had four coefficients, and so does the transform. Also note that, given the transform, we can reconstruct the image to any resolution by recursively adding and subtracting the detail coefficients from the lower-resolution versions.

Storing the wavelet transform of the image, rather than the image itself, has a number of advantages. One advantage of the wavelet transform is that often a large number of the detail coefficients turn out to be very small in magnitude, as in the larger example of Figure 2.1. Truncating, or removing, these small coefficients from the representation introduces only small errors in the reconstructed image, giving a form of "lossy" image compression. We will discuss this particular application of wavelets in Section 2.4, once we have presented the one-dimensional Haar basis functions.

2.2 One-dimensional Haar basis functions

In the previous section we treated one-dimensional images as sequences of coefficients. Alternatively, we can think of images as piecewise-constant functions on the half-open interval $[0, 1)$. (A *half-open interval* $[a, b)$ contains all values of x in the range $a \leq x < b$.) In this new treatment, we will use the concept of a "vector space" from linear algebra. (A refresher on linear algebra can be found in Appendix A.)

A *vector space V* is basically just a collection of "things" (called *vectors*, in this context) for which addition and scalar multiplication are defined. Thus, you can add two vectors, scale a vector by some constant, and so forth. (The full list of axioms can be found in Appendix A.1.)

Until now, we have been thinking of images as sequences of coefficients; let's instead think of them as functions. For example, we can consider a one-pixel image to be a function that is constant over the entire interval $[0, 1)$. Since addition and scalar multiplication of functions are well defined, we can then think of each constant function over the interval $[0, 1)$ as a vector, and we'll let V^0 denote the vector space of all such functions. Similarly, a two-pixel image is a function having two constant pieces over the intervals $[0, 1/2)$ and $[1/2, 1)$. We'll call the space containing all these functions V^1. If we continue in this manner, the space V^j will include all piecewise-constant functions defined on the interval $[0, 1)$ with constant pieces over each of 2^j equal-sized subintervals.

Resolution	Approximation	Detail Coefficients

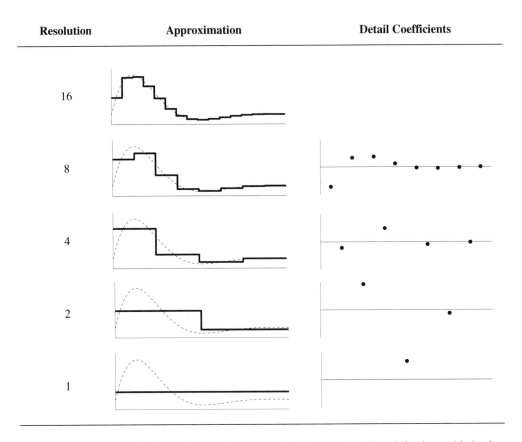

FIGURE 2.1   A sequence of decreasing-resolution approximations to a function (left) along with the detail coefficients required to recapture the finest approximation (right). Note that in regions where the true function is close to being flat, a piecewise-constant approximation is a good one, and so the corresponding detail coefficients are relatively small.

We can now think of every one-dimensional image with 2^j pixels as being an element, or vector, in V^j. Note that because these vectors are all functions defined on the unit interval, every vector in V^j is also contained in V^{j+1}. For example, we can always describe a piecewise-constant function with two intervals as a piecewise-constant function with four intervals, with each interval in the first function corresponding to a pair of intervals in the second. Thus, the spaces V^j are nested; that is,

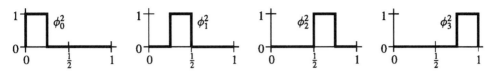

FIGURE 2.2 The box basis for V^2.

$$V^0 \subset V^1 \subset V^2 \subset \cdots$$

This nested set of spaces V^j is a necessary ingredient for the mathematical theory of multiresolution analysis, a topic we will consider more thoroughly in Chapter 7.

Now we need to define a basis for each vector space V^j. (A *basis* for a vector space is defined formally in Appendix A.2. Roughly speaking, a basis consists of a minimum set of vectors from which all other vectors in the vector space can be generated through linear combinations.) The basis functions for the spaces V^j are called *scaling functions* and are usually denoted by the symbol ϕ. A simple basis for V^j is given by the set of scaled and translated *box functions:*

$$\phi_i^j(x) := \phi(2^j x - i) \qquad i = 0, \dots, 2^j - 1$$

where

$$\phi(x) := \begin{cases} 1 & \text{for } 0 \leq x < 1 \\ 0 & \text{otherwise} \end{cases}$$

As an example, the four box functions forming a basis for V^2 are shown in Figure 2.2.

The *support* of a function refers to the region of the parameter domain over which the function is nonzero. For example, the support of $\phi_0^2(x)$ is $[0, 1/4)$. Functions that are supported over a bounded interval are said to have *compact support*. Note that all of the box functions are compactly supported.

The next step in building a multiresolution analysis is to choose an *inner product* defined on the vector spaces V^j (see Appendix A.3 for a formal definition of inner products). For our running example, the "standard" inner product will do quite well:

$$\langle f \mid g \rangle := \int_0^1 f(x)\, g(x)\, dx$$

Two vectors u and v are said to be *orthogonal* under a chosen inner product if $\langle u \mid v \rangle = 0$. We can now define a new vector space W^j as the *orthogonal complement* of V^j in V^{j+1}. In other words, W^j is the space of all functions in V^{j+1} that are orthogonal to all functions in V^j under the chosen inner product.

A collection of linearly independent functions $\psi_i^j(x)$ spanning W^j are called *wavelets*. These basis functions have two important properties:

1. The basis functions ψ_i^j of W^j, together with the basis functions ϕ_i^j of V^j, form a basis for V^{j+1}.

2. Every basis function ψ_i^j of W^j is orthogonal to every basis function ϕ_i^j of V^j under the chosen inner product.

Remark: Later, in Chapter 7, we'll look at ways in which the definitions of the complement spaces W^j and the wavelets ψ_i^j above can either be made more strict or more relaxed. For example, some authors refer to the functions defined as wavelets above as *pre-wavelets*, reserving the term *wavelets* for functions ψ_i^j that are orthogonal to each other as well. ■

Informally, we can think of the wavelets in W^j as a means of representing the parts of a function in V^{j+1} that cannot be represented in V^j. Thus, the detail coefficients of Section 2.1 are really coefficients of the wavelet basis functions.

The wavelets corresponding to the box basis are known as the *Haar wavelets*, given by

$$\psi_i^j(x) := \psi(2^j x - i) \qquad i = 0, \ldots, 2^j - 1$$

where

$$\psi(x) := \begin{cases} 1 & \text{for } 0 \le x < 1/2 \\ -1 & \text{for } 1/2 \le x < 1 \\ 0 & \text{otherwise} \end{cases}$$

Figure 2.3 shows the two Haar wavelets spanning W^1.

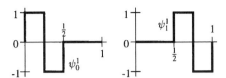

FIGURE 2.3 The Haar wavelets for W^1.

Before going on, let's run through our example from Section 2.1 again, but now applying these more sophisticated ideas. We begin by expressing our original image $\mathcal{I}(x)$ as a linear combination of the box basis functions in V^2:

$$\mathcal{I}(x) = c_0^2\, \phi_0^2(x) + c_1^2\, \phi_1^2(x) + c_2^2\, \phi_2^2(x) + c_3^2\, \phi_3^2(x)$$

A more graphical representation is

$$\mathcal{I}(x) = 9 \times$$

$$+ 7 \times$$

$$+ 3 \times$$

$$+ 5 \times$$

Note that the coefficients $c_0^2, \ldots, c_3^2$ are just the four original pixel values [9 7 3 5].

We can rewrite the expression for $\mathcal{I}(x)$ in terms of basis functions in V^1 and W^1, using pairwise averaging and differencing:

$$\mathcal{I}(x) = c_0^1\, \phi_0^1(x) + c_1^1\, \phi_1^1(x) + d_0^1\, \psi_0^1(x) + d_1^1\, \psi_1^1(x)$$

$$= 8 \times$$

$$+ 4 \times$$

$$+ 1 \times$$

$$+ -1 \times$$

These four coefficients should look familiar as well.

Finally, we'll rewrite $\mathcal{I}(x)$ as a sum of basis functions in V^0, W^0, and W^1:

$$\mathcal{I}(x) = c_0^0\, \phi_0^0(x) + d_0^0\, \psi_0^0(x) + d_0^1\, \psi_0^1(x) + d_1^1\, \psi_1^1(x)$$

$$= 6 \times$$

$$+ 2 \times$$

$$+ 1 \times$$

$$+ {-1} \times$$

Once again, these four coefficients are the Haar wavelet transform of the original image. The four functions shown above constitute the Haar basis for V^2. Instead of using the usual four box functions, we can use ϕ_0^0, ψ_0^0, ψ_0^1, and ψ_1^1 to represent the overall average, the broad detail, and the two types of finer detail possible in a function in V^2. The Haar basis for V^j with $j > 2$ includes these four functions as well as even narrower versions of the wavelet $\psi(x)$.

2.3   Orthogonality and normalization

The Haar basis possesses an important property known as *orthogonality*, which is not always shared by other wavelet bases. An orthogonal basis is one in which all of the basis functions, in this case ϕ_0^0, ψ_0^0, ψ_1^0, ψ_1^1, ..., are orthogonal to one another. Note that orthogonality is stronger than the requirement in the definition of wavelets that ψ_i^j be orthogonal only to all scaling functions at the same hierarchy level j.

Another property of some wavelet bases is *normalization*. A basis function $u(x)$ is normalized if $\langle u \mid u \rangle = 1$. We can normalize the Haar basis by replacing our earlier definitions with

$$\phi_i^j(x) := \sqrt{2^j}\, \phi(2^j x - i)$$
$$\psi_i^j(x) := \sqrt{2^j}\, \psi(2^j x - i)$$

where the constant factor $\sqrt{2^j}$ is chosen to satisfy $\langle u \mid u \rangle = 1$ for the standard inner product. With these modified definitions, the new normalized coefficients are obtained by dividing each old coefficient with superscript j by $\sqrt{2^j}$. Thus, in the example from the previous section, the unnormalized coefficients $[6\ 2\ 1\ -1]$ become the normalized coefficients

$$\left[6\ 2\ \frac{1}{\sqrt{2}}\ \frac{-1}{\sqrt{2}}\right]$$

As an alternative to first computing the unnormalized coefficients and then normalizing them afterwards, we can include normalization in the decomposition algorithm. The following two pseudocode procedures accomplish this normalized decomposition, transforming a set of coefficients in place:

procedure *Decomposition*(c: **array** $[1 \ .. \ 2^j]$ **of reals**)
   $c \leftarrow c/\sqrt{2^j}$ *(normalize input coefficients)*
   $g \leftarrow 2^j$
  while $g \geq 2$ **do**
    DecompositionStep($c[1 \ .. \ g]$)
    $g \leftarrow g/2$
  end while
end procedure

procedure *DecompositionStep*(c: **array** $[1 \ .. \ 2^j]$ **of reals**)
  for $i \leftarrow 1$ **to** $2^j/2$ **do**
    $c'[i] \leftarrow (c[2i-1] + c[2i])/\sqrt{2}$
    $c'[2^j/2 + i] \leftarrow (c[2i-1] - c[2i])/\sqrt{2}$
  end for
  $c \leftarrow c'$
end procedure

Of course, after we obtain a wavelet decomposition, we need to be able to reconstruct the original data. The following two pseudocode procedures do just that.

procedure *Reconstruction*(c: **array** $[1 \ .. \ 2^j]$ **of reals**)
   $g \leftarrow 2$
  while $g \leq 2^j$ **do**
    ReconstructionStep($c[1 \ .. \ g]$)
    $g \leftarrow 2g$
  end while
  $c \leftarrow c\sqrt{2^j}$ *(undo normalization)*
end procedure

```
procedure ReconstructionStep(c: array [1 .. 2^j] of reals)
    for i ← 1 to 2^j/2 do
        c'[2i − 1] ← (c[i] + c[2^j/2 + i])/√2
        c'[2i] ← (c[i] − c[2^j/2 + i])/√2
    end for
    c ← c'
end procedure
```

The above pseudocode procedures allow us to work with an *orthonormal* basis: one that is both orthogonal and normalized. As we will see in the next section, using an orthonormal basis turns out to be handy when compressing a function or an image.

2.4   Wavelet compression

The goal of compression is to express an initial set of data using some smaller set of data, either with or without loss of information. For instance, suppose we are given a function $f(x)$ expressed as a weighted sum of basis functions $u_1(x), \ldots, u_m(x)$:

$$f(x) = \sum_{i=1}^{m} c_i \, u_i(x)$$

The data set in this case consists of the coefficients $c_1, \ldots, c_m$. We would like to find a function approximating $f(x)$ but requiring fewer coefficients, perhaps by using a different basis. That is, given a user-specified error tolerance ε (for lossless compression, $\varepsilon = 0$), we are looking for

$$\hat{f}(x) = \sum_{i=1}^{\hat{m}} \hat{c}_i \, \hat{u}_i(x)$$

such that $\hat{m} < m$ and $\| f(x) - \hat{f}(x) \| \leq \varepsilon$ for some norm (see Appendix A.4 for more on norms). In general, one could attempt to construct a set of basis functions $\hat{u}_1, \ldots, \hat{u}_{\hat{m}}$ that would provide a good approximation with few coefficients. We will focus instead on the simpler problem of finding a good approximation in a fixed basis. Note that here and elsewhere in the book, when we discuss compression, we are concentrating on reducing the number of coefficients needed to represent a function—and not on the equally challenging problem of *encoding* and storing the necessary information in the fewest possible bits.

One form of the compression problem is to order the coefficients $c_1, \ldots, c_m$ so that for every $\hat{m} < m$, the first $\hat{m}$ elements of the sequence give the best approximation $\hat{f}(x)$ to $f(x)$ as measured in the L^2 norm. As we show here, the solution to this problem is straightforward if the basis is orthonormal, as is the case with the normalized Haar basis.

Let $\pi(i)$ be a permutation of $1, \ldots, m$ and let $\hat{f}(x)$ be a function that uses the coefficients corresponding to the first $\hat{m}$ numbers of the permutation $\pi(i)$:

$$\hat{f}(x) = \sum_{i=1}^{\hat{m}} c_{\pi(i)} \, u_{\pi(i)}$$

The square of the L^2 error in this approximation is given by

$$\left\| f(x) - \hat{f}(x) \right\|_2^2 = \left\langle f(x) - \hat{f}(x) \mid f(x) - \hat{f}(x) \right\rangle$$

$$= \left\langle \sum_{i=\hat{m}+1}^{m} c_{\pi(i)} \, u_{\pi(i)} \; \middle| \; \sum_{j=\hat{m}+1}^{m} c_{\pi(j)} \, u_{\pi(j)} \right\rangle$$

$$= \sum_{i=\hat{m}+1}^{m} \sum_{j=\hat{m}+1}^{m} c_{\pi(i)} \, c_{\pi(j)} \left\langle u_{\pi(i)} \mid u_{\pi(j)} \right\rangle$$

$$= \sum_{i=\hat{m}+1}^{m} (c_{\pi(i)})^2$$

The last step follows from the assumption that the basis is orthonormal, so $\langle u_i \mid u_j \rangle = \delta_{ij}$. The above result indicates that the square of the overall L^2 error is just the sum of the squares of all the coefficients we choose to leave out. We conclude that in order to minimize this error for any given $\hat{m}$, the best choice for $\pi(i)$ is the permutation that sorts the coefficients in order of decreasing magnitude; that is, $\pi(i)$ satisfies

$$\left| c_{\pi(1)} \right| \geq \cdots \geq \left| c_{\pi(m)} \right|$$

Figure 2.1 demonstrated how a one-dimensional function could be transformed into coefficients representing the function's overall average and various resolutions of detail. Now we

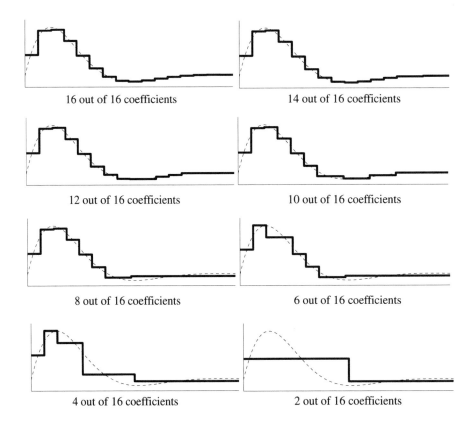

16 out of 16 coefficients   14 out of 16 coefficients

12 out of 16 coefficients   10 out of 16 coefficients

8 out of 16 coefficients   6 out of 16 coefficients

4 out of 16 coefficients   2 out of 16 coefficients

FIGURE 2.4   Coarse approximations to a function obtained using L^2 compression: detail coefficients are removed in order of increasing magnitude.

repeat the process, this time using normalized Haar basis functions. We can apply L^2 compression to the resulting coefficients simply by removing or ignoring the coefficients with smallest magnitude. By varying the amount of compression, we obtain a sequence of approximations to the original function, as shown in Figure 2.4.

IMAGE COMPRESSION

1. Two-dimensional Haar wavelet transforms — 2. Two-dimensional Haar basis functions — 3. Wavelet image compression — 4. Color images — 5. Summary

In preparation for image compression, we need to generalize Haar wavelets to two dimensions. First, we consider how to perform a wavelet decomposition of the pixel values in a two-dimensional image. We then describe the scaling functions and wavelets that form a two-dimensional wavelet basis. These tools will enable us to describe image compression as an application of wavelets.

3.1 Two-dimensional Haar wavelet transforms

There are two common ways in which wavelets can be used to transform the pixel values within an image. Each of these transformations is a two-dimensional generalization of the one-dimensional wavelet transform described in Section 2.1.

The first transform is called the *standard decomposition* [5]. To obtain the standard decomposition of an image, we first apply the one-dimensional wavelet transform to each row of pixel values. This operation gives us an average value along with detail coefficients for each row. Next, we treat these transformed rows as if they were themselves an image and apply the one-dimensional transform to each column. The resulting values are all detail coefficients

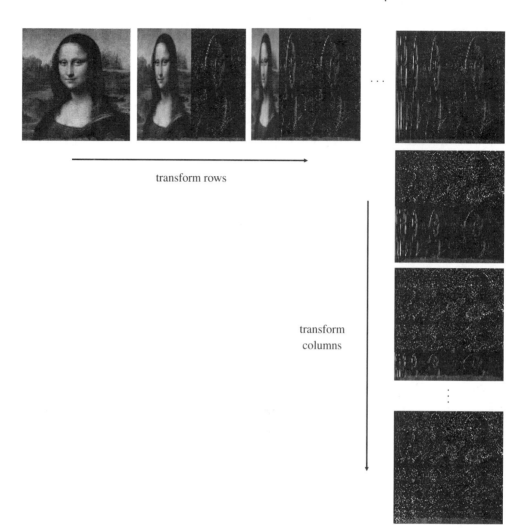

FIGURE 3.1   Standard decomposition of an image.

except for a single overall average coefficient. An algorithm to compute the standard decomposition is given below. Each step of its operation is illustrated in Figure 3.1.

> **procedure** *StandardDecomposition*(*c*: **array** $[1 \mathrel{..} 2^j, 1 \mathrel{..} 2^k]$ **of reals**)
>   **for** *row* ← 1 **to** 2^j **do**
>   *Decomposition*(*c*[*row*, $1 \mathrel{..} 2^k$])
>   **end for**
>   **for** *col* ← 1 **to** 2^k **do**

In the figure, labels read "transform rows" and "transform columns".

> $Decomposition(c[1 .. 2^j, col])$
> **end for**
> **end procedure**

The corresponding reconstruction algorithm simply reverses the steps performed during decomposition:

> **procedure** $StandardReconstruction(c:$ **array** $[1 .. 2^j, 1 .. 2^k]$ **of reals**)
>> **for** $col \leftarrow 1$ **to** 2^k **do**
>>> $Reconstruction(c[1 .. 2^j, col])$
>> **end for**
>> **for** $row \leftarrow 1$ **to** 2^j **do**
>>> $Reconstruction(c[row, 1 .. 2^k])$
>> **end for**
> **end procedure**

The second type of two-dimensional wavelet transform, called the *nonstandard decomposition* [5], alternates between operations on rows and columns. First, we perform one step of horizontal pairwise averaging and differencing on the pixel values in each row of the image. Next, we apply vertical pairwise averaging and differencing to each column of the result. To complete the transformation, we repeat this process recursively only on the quadrant containing averages in both directions. Figure 3.2 shows all the steps involved in the nonstandard decomposition procedure below.

> **procedure** $NonstandardDecomposition(c:$ **array** $[1 .. 2^j, 1 .. 2^j]$ **of reals**)
>> $c \leftarrow c/2^j$ *(normalize input coefficients)*
>> $g \leftarrow 2^j$
>> **while** $g \geq 2$ **do**
>>> **for** $row \leftarrow 1$ **to** g **do**
>>>> $DecompositionStep(c[row, 1 .. g])$
>>> **end for**
>>> **for** $col \leftarrow 1$ **to** g **do**
>>>> $DecompositionStep(c[1 .. g, col])$
>>> **end for**
>>> $g \leftarrow g/2$
>> **end while**
> **end procedure**

transform rows

transform
columns

FIGURE 3.2   Nonstandard decomposition of an image.

Here is the pseudocode to perform the nonstandard reconstruction:

> **procedure** *NonstandardReconstruction*(*c*: **array** $[1 .. 2^j, 1 .. 2^j]$ **of reals**)
> $g \leftarrow 2$
>   **while** $g \leq 2^j$ **do**
>   **for** $col \leftarrow 1$ **to** g **do**
>   *ReconstructionStep*($c[1 .. g, col]$)
>   **end for**
>   **for** $row \leftarrow 1$ **to** g **do**
>   *ReconstructionStep*($c[row, 1 .. g]$)
>   **end for**
> $g \leftarrow 2g$
>   **end while**
> $c \leftarrow 2^j c$ *(undo normalization)*
> **end procedure**

3.2   Two-dimensional Haar basis functions

The two methods of decomposing a two-dimensional image yield coefficients that correspond to two different sets of basis functions. The standard decomposition of an image gives coefficients for a basis formed by the *standard construction* of a two-dimensional basis. Similarly, the nonstandard decomposition gives coefficients for the *nonstandard construction* of basis functions [5].

The standard construction of a two-dimensional wavelet basis consists of all possible tensor products of one-dimensional basis functions. For example, when we start with the one-dimensional Haar basis for V^2, we get the two-dimensional basis for V^2 shown in Figure 3.3. Note that if we apply the standard construction to an orthonormal basis in one dimension, we get an orthonormal basis in two dimensions.

The nonstandard construction of a two-dimensional basis proceeds by first defining a two-dimensional scaling function,

$$\phi\phi(x, y) := \phi(x)\, \phi(y)$$

and three wavelet functions,

$$\phi\psi(x, y) := \phi(x)\, \psi(y)$$
$$\psi\phi(x, y) := \psi(x)\, \phi(y)$$
$$\psi\psi(x, y) := \psi(x)\, \psi(y)$$

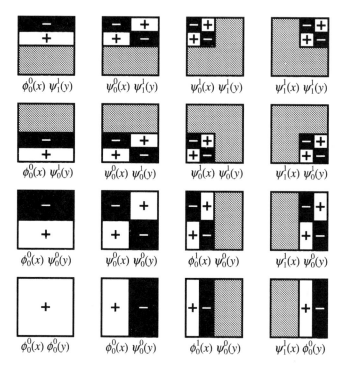

FIGURE 3.3   The standard construction of a two-dimensional Haar wavelet basis for V^2. In the unnormalized case, functions are +1 where plus signs appear, −1 where minus signs appear, and 0 in gray regions.

We now denote levels of scaling with a superscript j (as we did in the one-dimensional case) and horizontal and vertical translations with a pair of subscripts k and ℓ. The nonstandard basis consists of a single coarse scaling function $\phi\phi_{0,0}^0(x, y) := \phi\phi(x, y)$ along with scales and translates of the three wavelet functions $\phi\psi$, $\psi\phi$, and $\psi\psi$:

$$\phi\psi_{k\ell}^j(x, y) := 2^j \phi\psi(2^j x - k, 2^j y - \ell)$$
$$\psi\phi_{k\ell}^j(x, y) := 2^j \psi\phi(2^j x - k, 2^j y - \ell)$$
$$\psi\psi_{k\ell}^j(x, y) := 2^j \psi\psi(2^j x - k, 2^j y - \ell)$$

The constant 2^j normalizes the wavelets to give an orthonormal basis. The nonstandard construction results in the basis for V^2 shown in Figure 3.4.

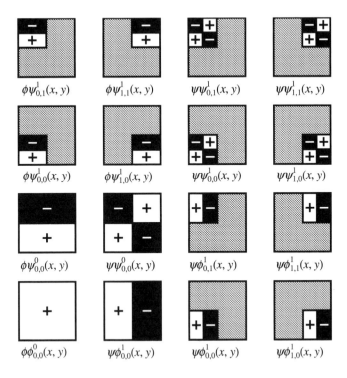

$\phi\psi^1_{0,1}(x,y)$ $\phi\psi^1_{1,1}(x,y)$ $\psi\psi^1_{0,1}(x,y)$ $\psi\psi^1_{1,1}(x,y)$

$\phi\psi^1_{0,0}(x,y)$ $\phi\psi^1_{1,0}(x,y)$ $\psi\psi^1_{0,0}(x,y)$ $\psi\psi^1_{1,0}(x,y)$

$\phi\psi^0_{0,0}(x,y)$ $\psi\psi^0_{0,0}(x,y)$ $\psi\phi^1_{0,1}(x,y)$ $\psi\phi^1_{1,1}(x,y)$

$\phi\phi^0_{0,0}(x,y)$ $\psi\phi^1_{0,0}(x,y)$ $\psi\phi^1_{0,0}(x,y)$ $\psi\phi^1_{1,0}(x,y)$

FIGURE 3.4 The nonstandard construction of a two-dimensional Haar wavelet basis for V^2.

We have presented both the standard and nonstandard approaches to wavelet transforms and basis functions because they each have advantages. The standard decomposition of an image is appealing because it can be accomplished simply by performing one-dimensional transforms on all the rows and then on all the columns. On the other hand, it is slightly more efficient to compute the nonstandard decomposition of an image. For an $m \times m$ image, the standard decomposition requires $4(m^2 - m)$ assignment operations, while the nonstandard decomposition requires only $\frac{8}{3}(m^2 - 1)$ assignment operations.

Another consideration is the support of each basis function, meaning the portion of each function's domain where that function is nonzero. All of the nonstandard Haar basis functions have square supports, while some of the standard basis functions have nonsquare supports. Depending upon the application, one of these choices may be preferable to the other.

3.3   Wavelet image compression

We defined compression in Section 2.4 as the representation of a function using fewer basis function coefficients than were originally given. The method we discussed for one-dimensional functions applies equally well to images, which we treat as the coefficients corresponding to a two-dimensional piecewise-constant basis. The approach presented here is only introductory; for a more complete treatment of wavelet image compression, see the article by DeVore et al. [27]. Once again, we note that we are dealing only with the transformation and quantization of coefficients and not with how they are encoded.

Wavelet image compression using the L^2 norm can be summarized in three steps:

1. Compute coefficients $c_1, \ldots, c_m$ representing an image in a normalized two-dimensional Haar basis.

2. Sort the coefficients in order of decreasing magnitude to produce the sequence $c_{\pi(1)}, \ldots, c_{\pi(m)}$.

3. Given an allowable L^2 error ε and starting with $\hat{m} = m$, find the smallest $\hat{m}$ for which

$$\sum_{i=\hat{m}+1}^{m} (c_{\pi(i)})^2 \leq \varepsilon^2$$

The first step is accomplished by applying either of the two-dimensional Haar wavelet transforms described in Section 3.1, being sure to use normalized basis functions. Any standard sorting technique will work for the second step, and any standard search can be used for the third step. However, for large images sorting becomes exceedingly slow. The pseudocode below outlines a more efficient method of accomplishing steps 2 and 3, which uses a binary search strategy to find a threshold τ below which coefficients can be truncated.

The procedure takes as input a one-dimensional array of coefficients c (with each coefficient corresponding to a two-dimensional basis function) and an error tolerance ε. For each guess at a threshold τ, the algorithm computes the square of the L^2 error that would result from discarding coefficients smaller in magnitude than τ. This squared error s is compared to ε^2 at each iteration to decide whether the binary search should continue in the upper or lower half of the current interval. The algorithm halts when the current interval is so narrow that the number of coefficients to be discarded no longer changes.

procedure *Compress*(c: **array** $[1 . . m]$ **of reals**; ε: **real**)
   $\tau_{min} \leftarrow \min\{\,|c[i]|\,\}$
   $\tau_{max} \leftarrow \max\{\,|c[i]|\,\}$
   do
      $\tau \leftarrow (\tau_{min} + \tau_{max})/2$
      $s \leftarrow 0$
      for $i \leftarrow 1$ **to** m **do**
         if $|c[i]| < \tau$ **then** $s \leftarrow s + |c[i]|^2$
      end for
      if $s < \varepsilon^2$ **then** $\tau_{min} \leftarrow \tau$ **else** $\tau_{max} \leftarrow \tau$
   until $\tau_{min} \approx \tau_{max}$
   for $i \leftarrow 1$ **to** m **do**
      if $|c[i]| < \tau$ **then** $c[i] \leftarrow 0$
   end for
end procedure

The binary search algorithm given above was used to produce the images in Figure 3.5. These images demonstrate the high compression ratios wavelets offer as well as some of the artifacts they introduce.

DeVore et al. [27] suggest that the L^1 norm is best suited to the task of image compression. Here is a pseudocode fragment for a "greedy" L^1 compression scheme, which works by accumulating in a two-dimensional array $\Delta[x, y]$ the error introduced by discarding a coefficient and checking whether this error has exceeded a user-specified threshold:

for each pixel (x, y) **do**
   $\Delta[x, y] \leftarrow 0$
end for
for $i \leftarrow 1$ **to** m **do**
   $\Delta' \leftarrow \Delta +$ error from discarding $c[i]$
   if $\Sigma_{x,y} |\Delta'[x, y]| < \varepsilon$ **then**
      $c[i] \leftarrow 0$
      $\Delta \leftarrow \Delta'$
   end if
end for

(a)   (b)   (c)   (d)

FIGURE 3.5 L^2 Haar wavelet image compression: The original image (a) can be represented using (b) 19% of its wavelet coefficients, with 5% relative L^2 error; (c) 3% of its coefficients, with 10% relative L^2 error; (d) 1% of its coefficients, with 15% relative L^2 error.

Note that this algorithm's results depend on the order in which coefficients are visited. One could imagine obtaining very different images (and degrees of compression) by varying the order—for example, by starting with the finest scale coefficients rather than the smallest coefficients. One could also imagine running a more sophisticated constrained optimization procedure to select the minimum number of coefficients subject to the error bound.

3.4   Color images

Up to now, our discussion of images has encompassed only single-component gray-scale images. However, wavelet transforms and compression techniques apply equally well to color images with three color components. For example, the pseudocode given above for L^2 compression will work for a color image if we perform a wavelet transform independently on each of the three color components of the image and treat the results as an array of vector-valued wavelet coefficients. Then, instead of using the absolute value of a scalar coefficient in the pseudocode, we use the L^2 norm (the usual vector magnitude) of a vector-valued coefficient. Examples of color images compressed using this algorithm are shown in Color Plate 1.

Furthermore, there are a number of ways in which color information can be used to obtain a wavelet transform that is even more sparse than those we have discussed. For instance, by first converting the pixel values in an image from *RGB* colors to *YIQ* colors [38], we can separate the luminance information (Y) from the chromatic information (I and Q). Once we compute the wavelet transform, we can apply an L^2 compression procedure to each of the components of the image separately. Since human perception is most sensitive to variation in Y and least sensitive to variation in Q, we can permit the compression scheme to tolerate a larger error in the Q component of the compressed image, thereby increasing the amount of compression. (The same principle allows U.S. color television signals to be broadcast with bandwidths of 4 MHz for Y, 1.5 MHz for I, and 0.6 MHz for Q.)

3.5 Summary

In this and the previous chapter, we have described Haar wavelets in one and two dimensions, as well as how they can be used to compress functions and images. The Haar basis is also useful for image editing and querying, as described in the next two chapters, as well as for global illumination, as described in Chapter 13.

The theoretical exposition of wavelets will continue in Chapters 6 and 7, which present the theory of subdivision curves and show how this theory can be used in developing a more complete mathematical framework for multiresolution analysis.

IMAGE EDITING

1. Multiresolution image data structures — 2. Image editing algorithm —
3. Boundary conditions — 4. Display and editing at fractional resolutions —
5. Image editing examples

In this chapter, we'll look at the application of Haar wavelets to image editing. Haar wavelets provide a foundation for representing images that may have different resolutions in different places.

In an image editing application, it is important for a user to be able to make sweeping changes at a coarse resolution, as well as to do fine detail work at high resolution. Ideally, the storage cost of the resulting image should be proportional only to the amount of detail present at each resolution; furthermore, the time complexity of an editing operation should be proportional only to the resolution at which the operation is performed. In addition, the user should be able to magnify the image to an arbitrary resolution and to work at any convenient scale.

The image painting and compositing system developed by Berman et al. [4], which we describe in this chapter, meets these goals in large part. The system makes use of a Haar wavelet decomposition of the image, which is stored in a sparse quadtree structure. This wavelet representation has many advantages over other techniques for maintaining multiple resolutions in an image. First, the wavelet representation of an image is concise, in that it contains the same number of wavelet coefficients as there are pixels in the image. Second, this representation supports compositing more efficiently than image pyramids. Finally, wavelets provide a very effective means of compressing images, as we saw in Chapter 3. By using a wavelet representation, an editing system can operate on compressed images directly, without having to uncompress and recompress, making the handling of large images more practical than with a pyramid-based scheme.

The multiresolution images produced by a system based on wavelets can be thought of as having different resolutions in different places. There are many applications of these multiresolution images, including

- interactive paint systems—an artist can work on a single image at various resolutions

- texture mapping—portions of a texture that will be seen up close can be defined in more detail

- satellite and other image databases—overlapping images created at different resolutions can be coalesced into a single multiresolution image

- "importance-driven" physical simulations [119]—solutions can be computed at different resolutions in different places

- virtual reality, hypermedia, and games—image detail can be explored interactively using essentially unlimited degrees of panning and zooming

- the "infinite desktop" user-interface metaphor [93]—a single virtual "desktop" with infinite resolution can be presented to the user

4.1   Multiresolution image data structures

Let $\mathcal{I}$ be a *multiresolution image*—that is, an image with different resolutions in different places. One could think of $\mathcal{I}$ as an image whose resolution varies adaptively according to need.

More formally, we will define $\mathcal{I}$ as a sum of piecewise-constant functions $\mathcal{I}^j$ at different resolutions $2^j \times 2^j$. In this sense, $\mathcal{I}$ can be thought of as having "infinite" resolution everywhere: a user zooming into $\mathcal{I}$ would see more detail as long as higher-resolution detail is present; once this resolution is exceeded, the pixels of the finest-resolution image would appear to become larger and larger constant-colored squares.

We store the multiresolution image $\mathcal{I}$ in a sparse quadtree structure T. The nodes of T have the usual correspondence with portions of the image: the root of T, at level 0, corresponds to the entire image; the root's four children, at level 1, correspond to the image's four quadrants; and so on, down the tree. Thus, each level j of quadtree T corresponds to a scaled version of multiresolution image $\mathcal{I}$ at resolution $2^j \times 2^j$. Note that by the usual convention,

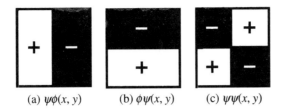

(a) $\psi\phi(x, y)$ (b) $\phi\psi(x, y)$ (c) $\psi\psi(x, y)$

FIGURE 4.1 The nonstandard Haar wavelets corresponding to the three types of detail represented by the coefficients d_1, d_2, and d_3.

"higher" levels of the quadtree correspond to lower-resolution versions of the image, and vice versa.

The quadtree is used to encode the nonstandard Haar wavelet decomposition we described in Sections 3.1 and 3.2. Each node of the quadtree contains the following information:

type *QuadTreeNode* = **record**
   d_1, d_2, d_3: *RGBA*
   τ: **real**
   child: **array** [1 . . 4] **of pointer to** *QuadTreeNode*
end record

The three d_i values in the *QuadTreeNode* structure are the detail coefficients: they describe how the colors of the children deviate from the color of the parent node. For a given node N, these d_i values are the coefficients of the three nonstandard Haar wavelets whose supports overlap the region of the image corresponding to N, with d_1, d_2, and d_3 the coefficients of $\psi\phi$, $\phi\psi$, and $\psi\psi$, respectively (see Figure 4.1). These coefficients allow us to reconstruct the *RGBA* colors of the four children, given the color of the parent, as described in Section 4.2.1.

The τ value represents the *transparency* of the node, initialized to 1. The τ values are used to optimize the painting and compositing algorithm, as explained later. The *child*[i] fields are pointers to the four children nodes. Some of these may be null. To optimize storage, the *child*[i] fields can alternatively be represented by a single pointer to an array of four children. We will later refer to the "alpha" component of a color c or detail coefficient d_i as $c.\alpha$ or $d_i.\alpha$.

Note that each node N of the tree corresponds to a particular region of the display. We will denote this region by *area* (N). The value *area* (N) is determined implicitly by the structure of the quadtree and the particular view, and does not need to be represented explicitly in N. Except when displaying at fractional levels (described in Section 4.4), there is a one-to-one correspondence between pixels on the display and nodes at some level j in the quadtree.

The quadtree itself is given by

type *QuadTree* = **record**
   c: *RGBA*
   root: **pointer to** *QuadTreeNode*
end record

The c value in the quadtree structure supplies the color of the root node; it corresponds to an average of all the colors in the image $\mathcal{I}$.

The quadtree is sparse in that it contains no leaves with detail coefficients that are all 0. Thus, the constant portions of the image at any particular resolution are represented implicitly. This convention allows us to support infinite resolutions in a finite structure. It also allows us to explicitly represent high-resolution details only where they actually appear in the image.

4.2   Image editing algorithm

Multiresolution painting is easy to implement. The main loop involves three steps: *Display*, *Painting*, and *Update*.

4.2.1   Display

An image at resolution $2^j \times 2^j$ is displayed by reconstructing the visible portion of the image from its two-dimensional wavelet decomposition. This reconstruction is accomplished by calling the following recursive *Display* routine once, passing it the root and color of the overall quadtree:

procedure *Display*(*N*: *QuadTreeNode*; *c*: *RGBA*)
   $c_1 \leftarrow c + N.d_1 + N.d_2 + N.d_3$
   $c_2 \leftarrow c - N.d_1 + N.d_2 - N.d_3$
   $c_3 \leftarrow c + N.d_1 - N.d_2 - N.d_3$
   $c_4 \leftarrow c - N.d_1 - N.d_2 + N.d_3$
   for $i \leftarrow 1$ **to 4 do**
     if *N* is a leaf **or** *N* is at level $j - 1$ **then**
       Draw c_i over the region *area*(*N.child*[*i*])
     else
       Display(*N.child*[*i*], c_i)
     end if

end for
end procedure

For clarity, the pseudocode above recurses to level $j - 1$ for the entire image; in reality, it should only recurse within the portion of the image that fits in the display window. Note that if m pixels are displayed, the entire display operation takes just $O(m)$ time. (More precisely, the operation requires $O(m + j)$ time; however, since $j \ll m$ in almost any practical situation, we will ignore this dependency on j in the analyses that follow.)

4.2.2 Painting

Painting is implemented by compositing the newly painted foreground buffer $\mathcal{F}$ with the background buffer $\mathcal{B}$ produced by *Display*, to create a new result image $\mathcal{R}$. The application described by Berman et al. [4] supports several binary compositing operations: "over," which places new paint wherever it is applied; "under," which places paint only where the background is transparent; and "in," which places paint only where the background is already painted. Their application also supports a unary "erase" operation, which removes paint from the background. The compositing algebra was originally described by Porter and Duff [96] and first described in the context of digital painting by Salesin and Barzel [104].

No special routines are required to implement painting itself. The only difference from ordinary painting is that in addition to the composited result $\mathcal{R}$, we must keep a separate copy of the foreground buffer $\mathcal{F}$, which contains all of the newly applied paint. This foreground buffer is necessary for updating the quadtree, as described in the next section. Ordinary painting proceeds until the user either changes the painting operation (for example, from "over" to "under") or changes the view by panning or zooming. Each of these actions triggers an update.

4.2.3 Update

The update operation is used to propagate the results of painting operations to the rest of the multiresolution image, as represented by the quadtree. The update involves two steps: *decomposition*, in which the changes are propagated to all higher levels of the quadtree, and *extrapolation*, in which the changes are propagated to all the lower levels. We will consider each of these in turn.

Let j be the level at which the user has been painting, and let $c_r(x, y)$ be the color of each modified pixel in the result image $\mathcal{R}$. A decomposition of the entire image is performed by calling the following *Decompose* function once, passing the root of the quadtree *T.root* as an argument, and storing the result in *T.c*:

function *Decompose*(*N*: *QuadTreeNode*)
   if *N* is at level *j* **then**
      return $c_r(x, y)$
   end if
   for $i \leftarrow 1$ **to** 4 **do**
      $c_i \leftarrow Decompose(N.child[i])$
   end for
   $N.d_1 \leftarrow (c_1 - c_2 + c_3 - c_4)/4$
   $N.d_2 \leftarrow (c_1 + c_2 - c_3 - c_4)/4$
   $N.d_3 \leftarrow (c_1 - c_2 - c_3 + c_4)/4$
   return $(c_1 + c_2 + c_3 + c_4)/4$
end function

For clarity, the pseudocode above assumes that the sparse quadtree *T* already contains all of the nodes corresponding to the pixels in the result image $\mathcal{R}$; however, if $\mathcal{R}$ has been painted at a higher resolution than the existing image, then new nodes may have to be allocated and added to *T* as part of the traversal. Furthermore, for efficiency, the *Decompose* function should be modified to recurse only in regions of the multiresolution image where changes have actually been made. Note that if the portion of the image being edited has *m* pixels, then the entire decomposition operation takes $O(m)$ time.

Extrapolation is a bit more complicated and depends on the particular compositing operation used. For binary painting operations, let $c_f(x, y)$ be the color of the foreground image $\mathcal{F}$ at each pixel (x, y), and let $c_f.\alpha(x, y)$ be the pixel's opacity value. For the "erase" operation, let $\delta(x, y)$ be the opacity of the eraser at each pixel. Extrapolation can then be performed by calling the following routine for the node *N* corresponding to each modified pixel (x, y) of the edited image:

procedure *Extrapolate*(*N*: *QuadTreeNode*)
   for $i \leftarrow 1$ **to** 3 **do**
      switch on the compositing operation
         case "over":
            $N.d_i \leftarrow N.d_i * (1 - c_f.\alpha(x, y))$
         case "under":
            $N.d_i \leftarrow N.d_i - N.d_i.\alpha * c_f(x, y)$
         case "in":
            $N.d_i \leftarrow N.d_i * (1 - c_f.\alpha(x, y)) + N.d_i.\alpha * c_f(x, y)$
         case "erase":
            $N.d_i \leftarrow N.d_i * (1 - \delta(x, y))$

      end switch
    end for
   if N is not a leaf **then**
    for $i \leftarrow 1$ **to** 4 **do**
      Extrapolate($N.child[i]$)
    end for
   end if
  end procedure

Note that the extrapolation procedure takes time proportional to the amount of detail that appears "below" the modified parts of the image. In order to optimize this operation, at least for the most common cases of painting "over" and erasing, we can use a form of lazy evaluation. First, observe that the two formulas for "over" and "erase" in the pseudocode above merely multiply the existing detail coefficients by some constant, which we will call $\tau(x, y)$. (For painting "over," $\tau(x, y) = 1 - c_f.\alpha(x, y)$; for "erase," $\tau(x, y) = 1 - \delta(x, y)$.) Thus, for these two operations, rather than calling the *Extrapolate* procedure for each node N, we can instead just multiply the value $N.\tau$ stored at the node by $\tau(x, y)$. Later, if and when the d_i values for a node N are actually required, they can be lazily updated by multiplying each $N.d_i$ with the τ values of all of the node's ancestors. This product is easily performed as part of the recursive evaluation.

This very simple form of lazy evaluation is a by-product of the underlying wavelet representation for the image, since the detail coefficients at higher resolutions depend only on the product of the opacities of all the paint applied and removed at lower resolutions. Any sort of lazy evaluation method would be much more complicated with image pyramids, since the high-resolution colors have a much more complex dependence on the colors of the paint applied and removed at lower resolutions.

Note that a kind of inexpensive "color correction" can be achieved by compositing a highly transparent constant-colored square over an entire image. Using the lazy evaluation, this kind of color correction can be performed on an arbitrarily high-resolution image in time proportional only to the resolution being displayed when the edit is performed.

4.3 Boundary conditions

Treating boundary conditions correctly introduces a slight complication to the update and display algorithms described in the sections above. The difficulty is that the *Decompose* function needs to have available to it the colors of the children of every node N that it traverses. However, some of these child nodes correspond to regions that are outside the boundary of the window in which the user has just painted and therefore are not directly available to the routine. The obvious solution is to store color information in addition to the detail coefficients at every node of the quadtree; however, this approach would significantly increase the storage requirements of the quadtree, as well as introduce the extra overhead of maintaining redundant representations. Instead, we keep a temporary auxiliary quadtree structure of just the colors necessary for the decomposition; this structure can be filled in during the *Display* operation at little extra cost. The size of this auxiliary structure is proportional to the number of pixels displayed.

4.4   Display and editing at fractional resolutions

So far, we have assumed a one-to-one correspondence between the nodes of the quadtree at level j and the pixels of the image at resolution $2^j \times 2^j$. Since the levels of the quadtree are discrete, this definition only provides for discrete levels of zooming in, which the resolution doubles at each level.

From a user-interface point of view, it would be better to be able to zoom in continuously on the multiresolution image being edited. A kind of *fractional-level* zooming can be defined by considering how the square region $area(N)$ corresponding to a given node N at level j in the quadtree would grow as a user zoomed in continuously from level j to $j + 1$ to $j + 2$. The size of $area(N)$ would increase exponentially from width 1 to 2 to 4 on the display. Thus, when displaying at a fractional level $j + \mu$, for some μ between 0 and 1, we would like $area(N)$ to have size $2^\mu \times 2^\mu$.

On workstations that provide antialiased polygon drawing, this fractional zooming is implemented quite simply by drawing each node N as a single square of the proper fractional size. On less expensive workstations that support only integer-sized polygons efficiently, a slightly less pleasing but still adequate display can be achieved by rounding each rendered square to the nearest pixel. In either case, the only change to the *Display* routine is to bottom out the recursion whenever N is at level $\lceil j + \mu - 1 \rceil$ instead of at level $j - 1$, and to let the region $area(child[i])$ correspond to the appropriate fractional size.

Of course, from a user's standpoint, if it is possible to display an image at any level $j + \mu$, then it should also be possible to edit it at that level. This fractional-level editing is also easy to support. To update the quadtree representation, we simply rescale the buffer of newly painted changes $\mathcal{F}$ to the size of the next higher integer level, as if the user had painted the changes at level $j + 1$; the scaling factor required is $2^{1-\mu}$. We can then perform the same update as before, starting from level $j + 1$.

4.5   Image editing examples

Color Plates 2 through 4 demonstrate the multiresolution image editing system with three examples. In Color Plate 2, the user magnifies an image of the Mona Lisa and paints on some eye shadow and lipstick at a high resolution. To add a glint in her eye, the user zooms in a little more. The retouched image is then displayed at the original resolution.

In Color Plate 3, the user paints a tree at multiple resolutions, using a variety of compositing operations. Most of the tree was painted at a coarse resolution. Then, in the first image, the user greatly magnifies the upper left corner of the tree and paints some leaves. Next, the user zooms out to a coarser scale and changes the color of the leaves, using an "in" brush that only paints where paint has previously been applied. In the third image, the user zooms out to a very coarse resolution and quickly roughs in the sky and grass, using an "under" brush that only deposits color where no paint already appears. Note that even though the sky color is applied coarsely, the new paint respects all of the high-resolution detail originally present in the image.

Finally, Color Plate 4 shows an example in which a single multiresolution image was created out of six successive images from the book *Powers of Ten* [87] by compositing the images together at different scales. In the book, each image magnifies the central portion of its predecessor by a factor of ten. In our multiresolution representation, these six images become a single image with a 10^5 range of scale. (Note that representing power-of-ten images in a power-of-two quadtree requires the fractional-level editing capability.) Color Plate 4 shows a close-up of the innermost detail and the same image after zooming out by a factor of 100,000. In Color Plate 4(c), the user retouches the low-resolution image using an "over" brush to give the impression of smog. This smog affects all of the closer views without eliminating any of the detail present, as demonstrated in the final image.

IMAGE QUERYING

*1. Image querying by content — 2. Developing a metric for image querying —
3. Image querying algorithm — 4. Image querying examples — 5. Extensions*

With the explosion of desktop publishing, the ubiquity of color scanners and digital media, and the advent of the World Wide Web, people now have easy access to tens of thousands of digital images. This trend is likely to continue, providing more and more people with access to increasingly large image databases.

As the size of these databases grows, traditional methods of finding a particular image break down. For example, while it is relatively easy for a person to quickly look over a few hundred thumbnail images to find a specific image query, it is much harder to locate that query among several thousand. Exhaustive search quickly breaks down as an effective strategy when the database becomes sufficiently large.

One commonly employed searching strategy is to index the image database with keywords. However, such an approach is also fraught with difficulties. First, it requires a person to manually tag all the images with keys, a time-consuming task. Second, as Niblack et al. point out [90], this keyword approach has the problem that some visual aspects are inherently difficult to describe, while others are equally well described in many different ways. In addition, it may be difficult for the user to guess which visual aspects have been indexed.

In this chapter, we explore an alternative strategy for searching an image database, in which the query is expressed either as a low-resolution image from a scanner or video camera or as a rough sketch of the image painted by the user. This basic approach to image querying has been referred to in a variety of ways, including "query by content" [90], "query by example" [59], "similarity retrieval" [68], and "sketch retrieval" [66]. Note that this type of content-based querying can also be applied in conjunction with keyword-based querying or any other existing approach.

Content-based querying has applications in many different domains, including graphic design, architecture, TV production, multimedia, ubiquitous computing, art history, geology, satellite image databases, and medical imaging. For example, a graphic designer may want to find an image that is stored on her own system using a painted query. She may also want to find out if a supplier of ultra–high-resolution digital images has a particular image in its database, using a low-resolution scanned query. In the realm of ubiquitous computing, a computer may need to find a given document in its database, given a video image of a page of that document scanned in from the real-world environment. In all of these applications, improving the technology for content-based querying is an important and acknowledged challenge.

Several factors make this problem difficult to solve. The "query" image is typically very different from the "target" image, so the retrieval method must allow for some distortions. If the query is scanned, it may suffer artifacts such as color shift, poor resolution, dithering effects, and misregistration. If the query is painted, it is limited by perceptual error in both shape and color, as well as by the artistic prowess and patience of the user. For these reasons, straightforward approaches such as L^1 or L^2 image metrics are not very effective in discriminating the target image from the rest of the database. In order to match such imperfect queries more effectively, a kind of "image querying metric" must be developed that accommodates these distortions and yet distinguishes the target image from the rest of the database. In addition, the retrieval should ideally be fast enough to handle databases with tens of thousands of images at interactive rates.

Jacobs et al. describe how a Haar wavelet decomposition of the query and database images can be used to match a content-based query both quickly and effectively [63]. Their *multiresolution image querying algorithm* greatly improves on the speed and robustness of other content-based image querying systems.

The input to their retrieval method is a sketched or scanned image, intended to be an approximation to the image being retrieved. Since the input is only approximate, the approach they take is to present the user with a small set of the most promising target images as output, rather than with a single "correct" match. Twenty images (the number of slides on a slide sheet) are about the most that can be scanned quickly and reliably by a user in search of the target.

In order to perform this ranking, Jacobs et al. define an *image querying metric* that makes use of truncated, quantized versions of the wavelet decompositions, called *signatures*.

The signatures contain only the most significant information about each image. The image querying metric essentially compares how many significant wavelet coefficients the query has in common with potential targets. The metric can be tuned, using statistical techniques, to discriminate most effectively for different types of content-based image querying, such as scanned or hand-painted images. Jacobs et al. also present a novel database organization for computing this metric extremely fast. (Their system processes a 128×128 image query on a database of 20,000 images in under $1/2$ second; by contrast, searching the same database using an L^1 metric takes over 14 minutes.)

5.1 Image querying by content

Previous approaches to content-based image querying have applied such properties as color histograms [121], texture analysis [65], and shape features such as circularity and major-axis orientation of regions in the image [48], as well as combinations of these techniques.

One of the most notable systems for querying by image content, called QBIC, was developed at IBM [90] and is now available commercially. The emphasis in QBIC is to allow a user to compose a query based on a variety of different visual attributes; for example, the user might specify a particular color composition ($x\%$ of color 1, $y\%$ of color 2, etc.), a particular texture, some shape features, and a rough sketch of dominant edges in the target image, along with relative weights for all of these attributes. The QBIC system also allows users to annotate database images by outlining key features to search for. By contrast, the emphasis of the multiresolution image querying algorithm is to search directly from a query image, without any further specifications from the user about the database images or about the particulars of the search itself.

The work of Hirata and Kato [59] is perhaps the most like the multiresolution querying technique in its style of user interaction. In their system, called query by visual example (QVE), edge extraction is performed on user queries. These edges are matched against those of the database images in a fairly complex process that allows for corresponding edges to be shifted or deformed with respect to each other.

A multiresolution approach to image querying offers many advantages over other techniques. The use of wavelets allows a query to be specified at any resolution (potentially different from that of the target); moreover, the running time and storage of the multiresolution method are independent of the resolutions of the database images. In addition, the signature information required by the algorithm can be extracted from a wavelet-compressed version of the image directly, allowing the signature database to be created conveniently from a set of compressed images. Finally, the algorithm is much simpler to implement and to use than most previous approaches.

5.2 Developing a metric for image querying

Consider the problem of computing the distance between a query image Q and a potential target image T. The most obvious metrics to consider are the L^1 and L^2 norms:

$$\|Q - T\|_1 = \sum_{i,j} |Q[i,j] - T[i,j]| \tag{5.1}$$

$$\|Q - T\|_2 = \left(\sum_{i,j} (Q[i,j] - T[i,j])^2 \right)^{1/2} \tag{5.2}$$

However, these metrics are not only expensive to compute, they are also fairly ineffective when it comes to matching an inexact query image in a large database of potential targets. For example, Jacobs et al. report that in their experience with scanned queries, the L^1 and L^2 error metrics rank their intended target image in the highest 1% of the database only 3% of the time. (This rank is computed by sorting the database according to its L^1 or L^2 distance from the query and evaluating the intended target's position in the sorted list.)

On the other hand, the target of the query image is almost always readily discernible to the human eye, despite such potential artifacts as color shifts, misregistration, dithering effects, and distortion (which, taken together, account for the relatively poor performance of the L^1 and L^2 metrics). The solution, it would seem, is to try to find an image metric that is "tuned" for the kind of errors present in image querying; that is, we would like a metric that counts primarily those types of differences that a human would use for discriminating images but that gives much less weight to the types of errors that a human would ignore for this task.

Since there is no obvious "correct" metric to use for image querying, we are faced with the problem of constructing one from scratch, using (informed) trial and error. The rest of this section describes the issues addressed by Jacobs et al. in developing their image querying metric.

5.2.1 A multiresolution approach

We want to construct an image metric that is fast to compute, that requires little storage for each database image, and that improves significantly upon the L^1 or L^2 metrics in discriminating the targets of inexact queries. For several reasons, Jacobs et al. hypothesized that a two-dimensional wavelet decomposition of the images would provide a good foundation on which to build such a metric:

- Wavelet decompositions allow for very good image approximation with just a few coefficients. This property has been exploited for lossy image compression, as described in Chapter 3. Typically, for compression, only the wavelet coefficients with the largest magnitude are used.

- Wavelet decompositions can be used to extract and encode edge information [79]. Edges are likely to be among the key features of a user-painted query.

- The coefficients of a wavelet decomposition provide information that is independent of the original image resolution. Thus, a wavelet-based scheme allows the resolutions of the query and the target to be effectively decoupled.

- Wavelet decompositions can be computed quickly and easily, requiring linear time in the size of the image and very little code.

5.2.2 Components of the metric

Given the choice to use a wavelet approach, there are a number of issues that still need to be addressed:

1. *Color space.* We need to choose a color space in which to represent the images and perform the decomposition. (The same issue arises for L^1 and L^2 image metrics.) Jacobs et al. tried a number of different color spaces: *RGB*, *HSV*, and *YIQ*. Ultimately, *YIQ* turned out to be the most effective of the three for their data.

2. *Wavelet type.* Haar wavelets are the fastest to compute and simplest to implement. In addition, user-painted queries tend to have large constant-colored regions, which are well represented by this basis. One drawback of the Haar basis for lossy compression is that it tends to produce blocky image artifacts for high compression rates. In an image querying application, however, the results of the decomposition are never viewed, so these artifacts are of no concern.

3. *Decomposition type.* We need to choose either a standard or nonstandard two-dimensional wavelet decomposition (see Sections 3.1 and 3.2). In the Haar basis the nonstandard basis functions are square, whereas the standard basis functions are rectangular. We would therefore expect the nonstandard basis to be better at identifying features that are about as wide as they are high and the standard basis to work best for images containing vertical and horizontal lines, or other rectangular features.

4. *Truncation.* For a 128×128 image, there are $128^2 = 16{,}384$ different wavelet coefficients for each color channel. Rather than using all of these coefficients in the metric, it is preferable to truncate the sequence, keeping only the coefficients with largest magnitude. This truncation both accelerates the search for a query and reduces storage for the database. Surprisingly, truncating the coefficients also appears to *improve* the discriminatory power of the metric, probably because it allows the metric to consider only the most significant features and to ignore any mismatches in the fine detail, which the user, most likely, would have been unable to accurately recreate. Jacobs et al. report that storing the 60 largest-magnitude coefficients in each channel worked best for painted queries, while 40 coefficients worked best for scanned queries.

5. *Quantization.* Like truncation, the quantization of each wavelet coefficient can serve several purposes: speeding the search, reducing the storage, and actually improving the discriminatory power of the metric. The quantized coefficients retain little or no data about the precise magnitudes of major features in the images; however, the mere presence or absence of such features appears to have more discriminatory power for image querying than the features' precise magnitudes. Quantizing each significant coefficient to just two levels—+1, representing large positive coefficients, or −1, representing large negative coefficients—works remarkably well. This simple classification scheme also allows for a very fast comparison algorithm, as discussed in Section 5.3.

6. *Normalization.* The normalization of the wavelet basis functions is related to the magnitude of the computed wavelet coefficients: as the amplitude of each basis function increases, the size of that basis function's corresponding coefficient decreases accordingly (see Section 2.3). Using a normalization factor that makes all wavelets orthonormal to each other has the effect of emphasizing differences mostly at coarser scales.

5.2.3   The image querying metric

In order to write down the resulting metric, we must introduce some notation. First, let us now think of Q and T as representing just a single color channel of the wavelet decomposition of the query and target images. Let $Q[0, 0]$ and $T[0, 0]$ be the scaling function coefficients corresponding to the overall average intensity of that color channel. Further, let $\hat{Q}[i, j]$ and $\hat{T}[i, j]$ represent the $[i, j]$-th *truncated, quantized wavelet coefficients* of Q and T; these values are −1, 0, or +1. For convenience, we will let $\hat{Q}[0, 0]$ and $\hat{T}[0, 0]$, which do not correspond to any wavelet coefficient, equal 0.

A suitable metric for image querying can then be written as

$$\|Q - \mathcal{T}\| = w_{0,0} \, |Q[0, 0] - \mathcal{T}[0, 0]| + \sum_{i,j} w_{i,j} \, |\hat{Q}[i, j] - \hat{\mathcal{T}}[i, j]|$$

which we can simplify in a number of ways.

First, the metric is just as effective if the difference between the wavelet coefficients $|\hat{Q}[i, j] - \hat{\mathcal{T}}[i, j]|$ is replaced by $(\hat{Q}[i, j] \neq \hat{\mathcal{T}}[i, j])$, where the expression $(a \neq b)$ is interpreted as 1 if $a \neq b$ and 0 otherwise. This expression will be faster to compute in the algorithm.

Second, we would like to group terms together into "buckets" so that only a small number of weights $w_{i,j}$ need to be determined experimentally. We group the terms according to the scale of the wavelet functions to which they correspond, using a simple bucketing function $bin(i, j)$, described in detail in Section 5.3.

Finally, in order to make the metric even faster to evaluate over many different target images, we only consider terms in which the query has a nonzero wavelet coefficient $\hat{Q}[i, j]$. A potential benefit of this approach is that it allows for a query without much detail to match a very detailed target image quite closely; however, it does not allow a detailed query to match a target that does not contain that same detail. Jacobs et al. felt that this asymmetry might better capture the form of most painted image queries. (Note that this last modification technically disqualifies the "metric" from being a metric at all, since metrics, by definition, are symmetric. Nevertheless, for lack of a better term, we will continue to use the word *metric* in the rest of this chapter.)

Thus, the final "L^q" image querying metric is given by

$$\|Q - \mathcal{T}\|_q := w_0 \, |Q[0, 0] - \mathcal{T}[0, 0]| \; + \sum_{i,j:\hat{Q}[i,j]\neq 0} w_{bin(i,j)} \left(\hat{Q}[i, j] \neq \hat{\mathcal{T}}[i, j] \right) \tag{5.3}$$

The weights w in equation (5.3) provide a convenient mechanism for tuning the metric to different databases and styles of image querying. The actual weights used by Jacobs et al. are given in Section 5.3.

5.2.4 Fast computation of the image querying metric

To actually compute the L^q metric over a database of images, it is generally quicker to count the number of *matching* $\hat{Q}$ and $\hat{\mathcal{T}}$ coefficients rather than *mismatching* coefficients, as we expect the vast majority of database images not to match the query image well at all. It is therefore convenient to rewrite the summation in equation (5.3) in terms of an "equality" operator $(a = b)$, which evaluates to 1 when $a = b$ and 0 otherwise. Using this operator, the summation

$$\sum_{i,j:\hat{Q}[i,j]\neq 0} w_{bin(i,j)}\left(\hat{Q}[i,j]\neq\hat{T}[i,j]\right)$$

in equation (5.3) can be rewritten as

$$\sum_{i,j:\hat{Q}[i,j]\neq 0} w_{bin(i,j)} - \sum_{i,j:\hat{Q}[i,j]\neq 0} w_{bin(i,j)}\left(\hat{Q}[i,j]=\hat{T}[i,j]\right)$$

Since the first sum in this expression is independent of $\hat{T}$, we can ignore it for the purposes of ranking the different target images. It therefore suffices to compute the expression

$$w_0\left|Q[0,0]-T[0,0]\right| - \sum_{i,j:\hat{Q}[i,j]\neq 0} w_{bin(i,j)}\left(\hat{Q}[i,j]=\hat{T}[i,j]\right)$$

$$(5.4)$$

This expression is just a weighted sum of the difference in the average color between Q and T and the number of stored wavelet coefficients of T whose indices and signs match those of Q.

5.3   Image querying algorithm

The final algorithm is a straightforward embodiment of the L^q metric as given in equation (5.4) applied to the problem of finding a given query in a large database of images. The complexity of the algorithm is linear in the number of database images.

At a high level, the algorithm can be described as follows. In a preprocessing step, we perform a standard two-dimensional Haar wavelet decomposition (Section 3.1) of every image in the database and store only the overall average color and the indices and signs of the m wavelet coefficients of largest magnitude. The indices for all of the database images are then organized into a single data structure that optimizes searching. Then, for each query image, we perform the same wavelet decomposition and again throw away all but the average color and the largest m coefficients. The score for each target image T is then computed by evaluating equation (5.4). The rest of Section 5.3 describes this algorithm in more detail.

5.3.1   Preprocessing step

The preprocessing step reduces each image in the database to a signature as illustrated in Figure 5.1. A standard two-dimensional Haar wavelet decomposition of an image is very simple to code. It involves a one-dimensional decomposition of each row of the image, followed by a one-dimensional decomposition of each column of the result, as described in Section 3.1.

After the decomposition process, the entry $T[0,0]$ is proportional to the average color of the overall image, while the other entries of T contain the wavelet coefficients. (These co-

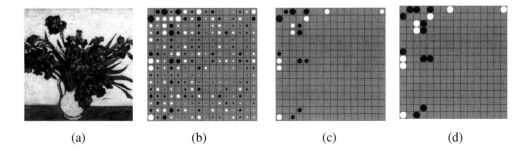

      (a)   (b)   (c)   (d)

FIGURE 5.1 Preprocessing steps: van Gogh's painting "Irises" (a) is first decomposed into wavelet coefficients (b). Next, all but the m largest-magnitude coefficients are truncated (c). Finally, the remaining coefficients are quantized (d). In the diagrams above, wavelet coefficients are represented by black and white dots. A dot's color (black or white) gives the sign (positive or negative) of the coefficient it represents. A dot's radius gives the magnitude of the coefficient.

efficients are sufficient for reconstructing the original image $\mathcal{T}$, although we will have no need to do so in this application.)

    Finally, we store only $\mathcal{T}[0, 0]$ and the indices and signs of the largest m wavelet coefficients of $\mathcal{T}$. To optimize the search process, the remaining m wavelet coefficients for *all* of the database images are organized into a set of six arrays, called the *search arrays*, with one array for every combination of sign (+ or −) and color channel (such as Y, I, and Q).

    For example, let Θ^c_+ denote the "positive" search array for the color channel c. Each element $\Theta^c_+[i, j]$ of this array contains a list of all images $\mathcal{T}$ having a large positive wavelet coefficient $\mathcal{T}[i, j]$ in color channel c. Similarly, each element $\Theta^c_-[i, j]$ of the "negative" search array points to a list of images with large negative coefficients in c. These six arrays are used to speed the search for a particular query, as described in the next subsection.

5.3.2 Querying

The querying step is straightforward. For a given query image Q, we perform the same wavelet decomposition described in the previous section. Again, we keep only the overall average color and the indices and signs of the largest m coefficients in each color channel. The signature of the query image is compared to the signatures in the database, as illustrated in Figure 5.2.

    To compute a score, we loop through each color channel c. We first compute the differences between the query's average intensity in that channel $Q^c[0, 0]$ and those of the database images. Next, for each of the m nonzero, truncated wavelet coefficients $\hat{Q}^c[i, j]$, we search through the list corresponding to those database images containing the same large-magnitude coefficient and sign, and update each of those image's scores accordingly:

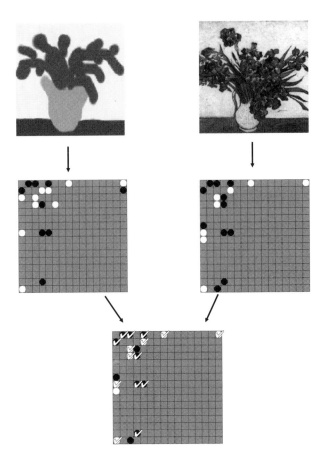

FIGURE 5.2   Comparing the signature of a query to the signature of a target image.

function *ScoreQuery*(Q: **array** $[0 .. r-1, 0 .. r-1]$ **of** *Color*; m: **integer**)
  DecomposeImage(Q)
  scores$[i] \leftarrow 0$ for all i
  for each color channel c **do**
    for each database image $\mathcal{T}$ **do**
      scores$[index(\mathcal{T})]$ += $w^c[0] * |Q^c[0, 0] - \mathcal{T}^c[0, 0]|$
    end for
    $\hat{Q} \leftarrow$ *TruncateCoefficients*(Q, m)

```
for each nonzero coefficient Q̂ᶜ[i, j] do
    if Q̂ᶜ[i, j] > 0 then
        list ← Θᶜ₊[i, j]
    else
        list ← Θᶜ₋[i, j]
    end if
    for each element ℓ of list do
        scores[index(ℓ)] − = wᶜ[bin(i, j)]
    end for
  end for
 end for
 return scores
end function
```

The function $bin(i, j)$ provides a way of grouping different coefficients into a small number of bins, with each bin weighted by some constant $w[b]$. Jacobs et al. discuss a statistical technique for finding the weights based on a set of training data. The size of their training set allowed them to determine 18 weights: 6 per color channel. Therefore, in their implementation, they use the function

$$bin(i, j) := \min\{\max\{level(i), level(j)\}, 5\}$$

where $level(i) = \lfloor \log_2 i \rfloor$. Table 5.1 gives the weights presented by Jacobs et al., which are tuned for their database of images using the YIQ color space and standard Haar decomposition. (All scaling function coefficients are reals between 0 and 1, so their differences tend to be smaller than the differences of the truncated, quantized wavelet coefficients. Thus, the weights on the scaling functions $w[0]$ have relatively large magnitudes because they generally multiply smaller quantities.)

As a final step, the image querying algorithm examines the list of scores, which may be positive or negative. The smallest scores are considered to be the closest matches. A "heapselect" algorithm [97] can be used to find the 20 closest matches in linear time.

5.4 Image querying examples

An implementation of the multiresolution image querying algorithm is illustrated in Color Plate 5. The user paints an image query in the large rectangular area on the left side of the application window. When the query is complete, the user presses the Match button. The system

TABLE 5.1   Weights used by Jacobs et al. in their multiresolution image querying metric.

b	Painted Queries			Scanned Queries		
	$w^Y[b]$	$w^I[b]$	$w^Q[b]$	$w^Y[b]$	$w^I[b]$	$w^Q[b]$
0	4.04	15.14	22.62	5.00	19.21	34.37
1	0.78	0.92	0.40	0.83	1.26	0.36
2	0.46	0.53	0.63	1.01	0.44	0.45
3	0.42	0.26	0.25	0.52	0.53	0.14
4	0.41	0.14	0.15	0.47	0.28	0.18
5	0.32	0.07	0.38	0.30	0.14	0.27

then tests the query against all the images in the database and displays the 20 top-ranked targets in the small windows on the right. (The highest-ranked target is displayed in the upper left, the second-highest target to its right, and so on, in row-major order.)

For convenience, the user may paint on a "canvas" of any aspect ratio. However, the application does not use this information in performing the match. Instead, the painted query is internally rescaled to a square aspect ratio and searched against a database in which all images have been similarly rescaled as a preprocess.

Color Plate 6(a) shows an example of a painted query, along with the L^q rank of its intended target in databases of 1093 and 20,558 images. Rather than painting a query, the user may also click on any of the displayed target images to serve as a subsequent query or use any stored image file as a query. Color Plate 6(b) shows an example of using a low-quality scanned image as a query, again displaying its L^q rank in the two databases.

The effectiveness of the L^q metric is demonstrated by the bar graph in Figure 5.3. The multiresolution image querying algorithm consistently retrieves intended targets more frequently than three other content-based metrics: the L^1 metric over 128×128 images, the L^1 metric over 8×8 low-resolution images, and the color histogram matching technique described by Jacobs et al.

As shown in Figure 5.4, the L^q metric also requires significantly less time to compute than other image querying metrics. In fact, because the retrieval time is so fast (under $1/2$ second in a database of 20,000 images), the image querying application described by Jacobs et al. supports an "interactive" mode, in which the 20 top-ranked target images are updated whenever the user pauses for a half-second or more. Color Plate 7 shows the progression of an interactive query, along with the actual time at which each snapshot was taken and the L^q rank of the intended target at that moment in the two different databases.

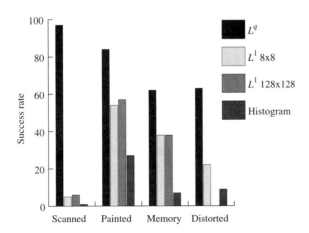

FIGURE 5.3   For each type of query (scanned, painted, painted from memory, and digitally distorted originals), the bars show what percentage of queries resulted in the correct target being ranked among the top 1% of images in a database of 1093 images by each metric.

While using interactive querying, users typically sketch all the information they know about an image in a minute or less, whether they are looking at a thumbnail image or painting from memory. In most cases, the query succeeds within this short time. If the query fails to bring up the intended target within a minute or so, users will typically try adding some random details, which sometimes help in bringing up the image. If this tactic fails, users will simply give up and, in a real system, would presumably fall back on some other method of searching for the image—or else use content-based image querying in conjunction with some other technique, such as a keyword search.

There are two benefits associated with painting queries interactively. First, the time to retrieve an image is generally reduced because the user simply paints until the target image appears, rather than painting until the query image seems finished. Second, the interactive mode may subtly help "train" the user to find images more efficiently, because the application is always providing feedback about the relative effectiveness of an unfinished query while it is being painted.

5.5   Extensions

The algorithm we have described is extremely fast, requires only a small amount of data to be stored for each target image, and is remarkably effective. It is also fairly easy to understand

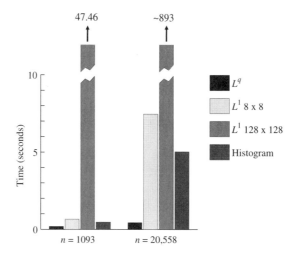

FIGURE 5.4   Average times (in seconds) to match a single query in databases of 1093 and 20,558 images using different metrics.

and implement. Finally, its parameters can be tuned for a given database or type of query image.

Here are some other ways in which the image querying algorithm might be extended:

- *Aspect ratio.* Currently, users can choose an aspect ratio for their query; however, this aspect ratio is not used in the search itself. It would be straightforward to add an extra term to the image querying metric for the similarity of aspect ratio. The weight for this term could be found experimentally at the same time the other weights are computed.

- *Perceptually based spaces.* It would be interesting to try using a perceptually uniform color space, such as CIE LUV or TekHVC [38], to see if it improves the effectiveness of the metric. In the same vein, it may help to compute differences on logarithmically scaled intensities, which is closer to the way intensity is perceived [57].

- *Image clusters.* Images in a large database appear to be "clustered" in terms of their proximity under the image querying metric. For example, using a portrait as a query image in a large image database selects portraits almost exclusively as targets. By contrast, using a planet image pulls up other planets. It would be interesting to perform some statistical clustering on the database and then show the user some representative images from the center of each cluster. These could be used either as querying keys, or merely as a way of providing an overview of the contents of the database.

- *Multiple metrics.* Experience with a multiresolution image querying system reveals that a good query will bring up the target image, no matter which color space and decomposition method (standard or nonstandard) is used. However, the false matches found in these different spaces all tend to be very different. Thus, it may be possible to develop a more effective method by combining the results of searching in different color spaces and decomposition types, perhaps taking the average of the ranks in the different spaces (or, alternatively, the worst of the ranks) as the rank chosen by the overall metric.

- *Affine transformation and partial queries.* A very interesting (and more difficult) direction for future research is to begin exploring methods for handling general affine transformations of the query image or for searching on partial queries. The "shiftable transforms," described by Simoncelli et al. [115], which allow for multiresolution transforms with translational, rotational, and scale invariance, may be helpful in these respects. Another idea for specifying partial queries would be to make use of the alpha channel of the query for specifying the portions of the query and target images over which the L^q metric should be computed.

- *Video querying.* The multiresolution image querying algorithm may well extend to the problem of searching for a given frame in a video sequence. The simplest solution would be to consider each frame of the video as a separate image in the database and to apply the image querying method directly. A more interesting solution would be to explore using a three-dimensional multiresolution decomposition of the video sequence, perhaps combined with some form of motion compensation, in order to take better advantage of the extra coherence among the video frames.

II

CURVES

6

SUBDIVISION CURVES

*1. Uniform subdivision — 2. Nonuniform subdivision — 3. Evaluation masks —
4. Nested spaces and refinable scaling functions*

The piecewise-constant spaces underlying Haar wavelet analysis are not sufficiently general for many applications. The functions of interest in curve and surface design, for instance, must at least be continuous, and they often must be once- or twice-differentiable. It is therefore necessary to extend our discussion to more general types of function spaces. However, this generalization must be done with considerable care, since very few function spaces admit hierarchical decompositions.

In this chapter and the next, we show that the only functions that can be hierarchically decomposed are those that can be generated through a simple process known as *recursive subdivision* (*subdivision* for short). Although not all functions can be defined in this way, subdivision can be used to create a surprisingly wide range of functions, including the piecewise-constant functions we considered in the context of Haar wavelets. Subdivision can also be used to create uniform and nonuniform B-splines, as well as functions that have no analytic form, such as the wavelets for which Daubechies is now famous. In the current chapter we develop the basic notions of subdivision for one-dimensional functions and parametric curves. Chapter 7 goes on to describe how these functions can be decomposed hierarchically.

The deep connection between subdivision and wavelets leads us to believe that subdivision methods will become increasingly important in the near future. Unfortunately, as of this writing we know of no text that emphasizes the practical aspects of subdivision, so we have attempted to provide such a treatment in this chapter. As a result, some of the material presented here is not central to the development of wavelets, but rather is included for the sake of completeness. In particular, Section 6.3 on the exact evaluation of functions defined through subdivision provides information that is useful for creating accurate renderings of subdivision curves and surfaces.

6.1   Uniform subdivision

The basic idea behind recursive subdivision is to create a function by repeatedly refining an initial piecewise-linear function $f^0(x)$ to produce a sequence of increasingly detailed functions $f^1(x), f^2(x), \ldots$ that converge to a limit function

$$f(x) := \lim_{j \to \infty} f^j(x)$$

The first subdivision scheme, illustrated in Figure 6.1, was introduced by Chaikin [11] in 1974. *Chaikin's algorithm*, like many other subdivision schemes, can be thought of as a "corner cutting" procedure to successively smooth the initial polygonal function into the final curve.

More concretely, let $f^0(x)$ be a piecewise-linear function with vertices at the integers. In general, the function $f^j(x)$ will be a piecewise-linear function with vertices at the *dyadic points* $i/2^j$. For most schemes, including Chaikin's algorithm, the values of $f^j(x)$ at its vertices are computed very simply as follows:

$$f^j\left(i/2^j\right) = \sum_k r_k \, f^{j-1}\left((i+k)/2^j\right)$$

$$(6.1)$$

The sequence $r = (\ldots, r_{-1}, r_0, r_1, \ldots)$ is called the *averaging mask* of the scheme. For Chaikin's algorithm, the averaging mask is $r = (r_0, r_1) = \frac{1}{2}(1, 1)$. Such a scheme is called a *uniform subdivision scheme* because the same mask is used everywhere along the curve (that is, r does not depend on i), and it is called *stationary* because the same mask is used on each iteration of subdivision (that is, r does not depend on j). Nonuniform schemes will be discussed in the next section; very little is known, however, about nonstationary schemes.

An easy way to implement recursive subdivision is to represent each function $f^j(x)$ by a list of the values $(\ldots, c^j_{-1}, c^j_0, c^j_1, \ldots)$ the function takes on at its vertices, where $c^j_i = f^j(i/2^j)$. Notice that the values of $f^{j-1}(x)$ required by equation (6.1) to define $f^j(x)$ consist

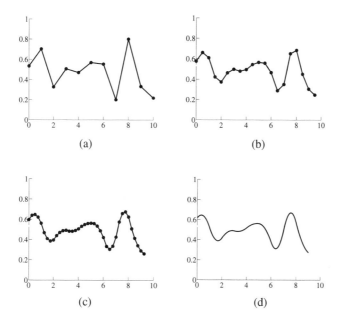

FIGURE 6.1 Chaikin's algorithm for a function: (a) the initial function $f^0(x)$; (b) the function $f^1(x)$; (c) the function $f^2(x)$; (d) the limit function $f(x)$.

of values at the vertices (when $i + k$ is even) and values at the midpoints (when $i + k$ is odd). To simplify the implementation of subdivision, it is convenient to separate the evaluation of equation (6.1) into two steps: the *splitting step*, which explicitly introduces midpoints, and the *averaging step*, which computes the weighted averages indicated by the equation. All subdivision schemes share the splitting step; they differ only in the averaging step.

The splitting step computes an auxiliary sequence $(\ldots, \overset{\circ}{c}{}^{j}_{-1}, \overset{\circ}{c}{}^{j}_{0}, \overset{\circ}{c}{}^{j}_{1}, \ldots)$ by introducing midpoints, as shown in Figure 6.2. Mathematically speaking, the splitting step is written as follows:

$$\overset{\circ}{c}{}^{j}_{2i} := c^{j-1}_i$$
$$\overset{\circ}{c}{}^{j}_{2i+1} := \frac{1}{2}\left(c^{j-1}_i + c^{j-1}_{i+1}\right)$$

As a result of this definition, $\overset{\circ}{c}{}^{j}_i = f^{j-1}(i/2^j)$ and, therefore, the averaging step is simply

$$c^j_i = \sum_k r_k \, \overset{\circ}{c}{}^{j}_{i+k}$$

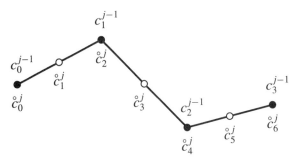

FIGURE 6.2   The splitting step transforms a sequence of vertices into a sequence of vertices and midpoints.

In practice we hardly ever have infinite sequences, so boundary conditions must be introduced. A common rule is to restrict the summation over k to the nonzero entries of r and to only compute values c_i^j for which all necessary values $\overset{\circ}{c}_i^j$ have been computed. A more principled way to deal with boundaries is to use nonuniform subdivision, as described in Section 6.2.

Like all subdivision schemes, Chaikin's algorithm can be applied to parametric curves, as well as to functions, simply by applying the splitting and averaging steps independently to the x- and y-coordinates of an initial polygon. This process is illustrated in Figure 6.3. The vertices of the initial polygon are called *control points*, and the polygon itself is called a *control polygon*.

It is not at all obvious how to determine properties such as continuity and differentiability of the functions created by an arbitrary mask r. Nonetheless, in 1975, Riesenfeld was able to show that the curves generated by Chaikin's algorithm are in fact uniform quadratic B-splines [103]. Some time later, Lane and Riesenfeld showed that Chaikin's algorithm could be generalized to produce uniform B-splines of *any* degree by using masks whose entries come from Pascal's triangle [69]. In particular, they showed that B-splines of degree $n + 1$ are generated by the averaging mask whose entries are

$$\frac{1}{2^n}\left(\binom{n}{0}, \binom{n}{1}, \ldots, \binom{n}{n}\right)$$

Two particularly important cases in computer graphics applications are linear B-splines, defined by the *identity mask* $r = (r_0) = (1)$, and cubic B-splines, defined by the averaging mask $r = (r_{-1}, r_0, r_1) = \frac{1}{4}(1, 2, 1)$. Note that the identity mask used for linear B-splines just leaves the

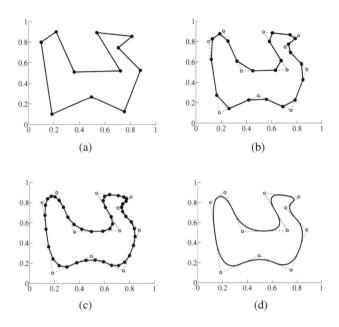

FIGURE 6.3 Chaikin's algorithm for a closed parametric curve: (a) the initial curve $\gamma^0(t)$; (b) the curve $\gamma^1(t)$; (c) the curve $\gamma^2(t)$; (d) the limit curve $\gamma(t) := \gamma^\infty(t)$.

vertices where they are, which is correct since performing the splitting step alone results in piecewise-linear interpolation.

Not all masks lead to familiar types of functions. For instance, the averaging mask $(r_0, r_1) = \frac{1}{2}(1 + \sqrt{3}, 1 - \sqrt{3})$ generates curves that are fractal-like in that they are nowhere differentiable, as shown in Figure 6.4. Although the usefulness of such a scheme may seem questionable now, this particular mask will come up again in the next chapter when we discuss a wavelet basis constructed by Daubechies using subdivision [24].

Aside from linear B-spline subdivision, the schemes we have mentioned so far are called *approximating schemes* since the limit function does not in general interpolate any of the initial vertices. It is possible to generate *interpolating schemes* by changing the averaging step so that a value, once computed, is never changed by local averaging. The change to the computation is straightforward:

$$c_i^j = \begin{cases} \overset{\circ}{c_i^j} & \text{if } i \text{ is even} \\ \sum_k r_k \overset{\circ}{c}_{i+k}^j & \text{if } i \text{ is odd} \end{cases}$$

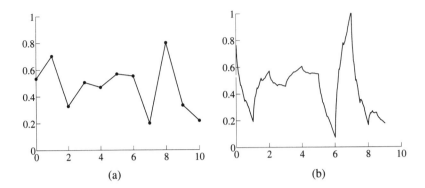

FIGURE 6.4   The Daubechies subdivision scheme: (a) the initial function $f^0(x)$; (b) the limit function $f(x)$.

The first smooth interpolating scheme was introduced by Dyn, Levin, and Gregory [32] (we'll refer to this algorithm as the *DLG scheme*). The averaging mask for the DLG scheme is $(r_{-2}, r_{-1}, r_0, r_1, r_2) = \frac{1}{16}(-2, 5, 10, 5, -2)$. An example of a parametric curve generated by the DLG scheme is shown in Figure 6.5.

6.2   Nonuniform subdivision

As mentioned earlier, the schemes in the previous section are called uniform or shift-invariant because the same mask is used everywhere on the curve. As we will see in chapters to come, for some applications it is more natural to allow the mask to vary along the curve. For example, the B-spline curve editing application presented in Chapter 8 uses *nonuniform* masks to create cubic B-splines that interpolate the first and last points of an open control polygon.

The general form of the averaging step for a nonuniform scheme is

$$c_i^j = \sum_k r_{i,k}^j \overset{\circ}{c}_{i+k}^j$$

$$\tag{6.2}$$

Remark: Strictly speaking, equation (6.2) describes a nonuniform, *nonstationary* scheme because r depends on both i and j. Very little is known about nonstationary subdivision, so we will assume that the scheme is stationary, at least in the sense that there is some critical value j^* such that for $j > j^*$ the entries of r are independent of j. (If this sounds confusing, you could just pretend that nonstationary schemes are well understood.) ■

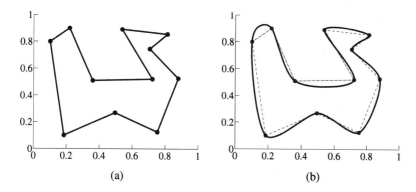

FIGURE 6.5   The DLG interpolating scheme for a closed parametric curve: (a) the control polygon $\gamma^0(t)$; (b) the limit curve $\gamma(t)$.

Before continuing the development of subdivision, we can spare ourselves some rather messy equations by writing equation (6.2) more compactly as a product of matrices. If we let c^j and $\mathring{c}^j$ denote the column matrices whose i-th rows contain c_i^j and $\mathring{c}_i^j$, respectively, and let R^j denote a matrix whose (i,k)-th entry is $r_{i,k}^j$, then equation (6.2) becomes

$$c^j = R^j \mathring{c}^j$$

The i-th row of R^j is the mask to use when computing the i-th entry of c^j, thus a uniform subdivision scheme is one in which the rows of R^j are shifted versions of one another. For instance, the averaging matrix for uniform cubic B-splines has the form

$$R^j = \frac{1}{4}\begin{bmatrix} 1 & 2 & 1 & & \\ & 1 & 2 & 1 & \\ & & 1 & 2 & 1 \\ & & & & \ddots \end{bmatrix}$$

$$\text{(6.3)}$$

where blank locations in the matrix indicate zeros.

Endpoint-interpolating B-splines will be treated in greater detail in Chapter 7, but we'll illustrate nonuniform subdivision with one example here. To describe cubic endpoint-interpolating B-splines, the first and last few rows of the averaging matrix in equation (6.3) must be modified to account for the endpoints. The rich theory of nonuniform B-spline knot insertion [35] can be used to show that for $j > 1$ (j must be large enough that the two endpoints do not affect each other), the first few rows of R^j are

$$\mathbf{R}^j = \frac{1}{4}\begin{bmatrix} 4 & & & & & & \\ & 4 & & & & & \\ & & 2 & 2 & & & \\ & & & \frac{3}{2} & \frac{3}{2} & 1 & \\ & & & & 1 & 2 & 1 \\ & & & & & 1 & 2 & 1 \\ & & & & & & & \ddots \end{bmatrix}$$

The rows in the middle of $\mathbf{R}^j$ are shifted versions of the uniform subdivision mask $\frac{1}{4}(1, 2, 1)$, and, by symmetry arguments, the last few rows are reversed versions of the first few rows.

A slightly simpler nonuniform scheme that is qualitatively similar to endpoint-interpolating B-splines was developed by Hoppe et al. [60]. For this scheme, only the first and last rows are modified:

$$\mathbf{R}^j = \frac{1}{4}\begin{bmatrix} 4 & & & & & \\ 1 & 2 & 1 & & & \\ & 1 & 2 & 1 & & \\ & & 1 & 2 & 1 & \\ & & & \ddots & & \\ & & & 1 & 2 & 1 \\ & & & & & 4 \end{bmatrix}$$

The limit curves resulting from this scheme are similar to endpoint-interpolating cubic B-splines in the following respects:

- Away from their ends, the limit curves are piecewise-cubic curves with two continuous derivatives.

- The limit curves interpolate their first and last control points.

- The limit curves are tangent to the first and last line segments of their control polygons.

6.3 Evaluation masks

As mentioned at the beginning of the chapter, this section is included for completeness in the treatment of subdivision. It can safely be skipped by readers interested only in the use of subdivision for creating wavelets.

Dealing with subdivision curves may at first appear to be rather unwieldy because, in general, they cannot be written in closed form. The lack of a closed form means, for instance, that a function or curve cannot be evaluated at an arbitrary parameter value simply by plugging a number into a formula. Differentiating the function or curve also appears to be problematic. We show in the remainder of this section that—somewhat surprisingly—functions defined through subdivision can be *exactly* evaluated at an arbitrarily dense set of points. Evaluation is accomplished by taking weighted averages according to an *evaluation mask* specific to the subdivision procedure. Moreover, if the subdivision scheme is smooth, derivatives of the function can also be computed exactly using *derivative masks*.

While the ability to compute exact values and derivatives is helpful for generating accurate plots of curves, it is even more essential for creating accurately shaded images of surfaces. Although we won't apply subdivision to surfaces until Chapter 10, the concepts behind evaluation and derivative masks are best developed in the simpler context of functions and curves.

Exact evaluation is based on the observation that vertices can be tracked through the subdivision process. For concreteness, let's consider the situation for uniform cubic B-spline subdivision—that is, subdivision based on the averaging mask $\frac{1}{4}(1, 2, 1)$. As shown in Figure 6.6, a vertex c^0 of f^0 can be associated naturally with a sequence of vertices $\{c^1, c^2, \dots\}$ of $\{f^1, f^2, \dots\}$, respectively. After an infinite number of subdivisions, c^0 achieves its limit position c^∞. We now show that c^∞ can be computed directly and exactly from the initial vertices.

The key observation in determining c^∞ is that in each step of subdivision, the position of c^j and its immediate neighbors can be determined from c^{j-1} and its neighbors [30, 54]. In particular, if we use c^j_- and c^j_+ to denote the left and right neighbors of c^j, then the splitting and averaging steps for cubic B-splines can be combined to give

$$c^j_- = \frac{c^{j-1}_- + c^{j-1}}{2}$$

$$c^j = \frac{c^{j-1}_- + 6\,c^{j-1} + c^{j-1}_+}{8}$$

$$c^j_+ = \frac{c^{j-1} + c^{j-1}_+}{2}$$

In matrix form, these equations become

$$\begin{bmatrix} c^j_- \\ c^j \\ c^j_+ \end{bmatrix} = \frac{1}{8} \underbrace{\begin{bmatrix} 4 & 4 & 0 \\ 1 & 6 & 1 \\ 0 & 4 & 4 \end{bmatrix}}_{L} \begin{bmatrix} c^{j-1}_- \\ c^{j-1} \\ c^{j-1}_+ \end{bmatrix}$$

(6.4)

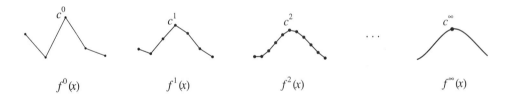

FIGURE 6.6   The tracking of a vertex c^0 of f^0 through the subdivision process.

The matrix L is known as the *local subdivision matrix* for the scheme. All schemes possess local subdivision matrices, but they're not always 3×3 matrices. The local subdivision matrix for Chaikin's algorithm, for instance, is

$$L = \frac{1}{4}\begin{bmatrix} 3 & 1 \\ 1 & 3 \end{bmatrix}$$

For stationary schemes, such as Chaikin's algorithm and others that generate uniform B-splines, the same subdivision matrix is used at each step of subdivision. Thus, vertices at level j can be computed directly by multiplying a small set of vertices at level 0 by the local subdivision matrix j times. This procedure holds all the way to the limit:

$$\begin{bmatrix} c_-^\infty \\ c^\infty \\ c_+^\infty \end{bmatrix} = \lim_{j \to \infty} (L)^j \begin{bmatrix} c_-^0 \\ c^0 \\ c_+^0 \end{bmatrix} \tag{6.5}$$

(Parentheses are used around L in the equation above to reinforce the fact that j here is being used as an exponent rather than a superscript.)

The *eigenstructure* of a square matrix governs what happens to a vector when it is multiplied by increasing powers of the matrix [47] (see Appendix A.5 for a brief review of eigenanalysis). Analysis of stationary subdivision schemes therefore reduces to eigenanalysis of the local subdivision matrix. One reason that so little is known about nonstationary schemes is that there are no analytical tools like eigenanalysis to appeal to when the local subdivision matrices depend on j.

To determine what happens in the limit of an infinite number of steps of stationary subdivision, it is useful to write the column vector $[c_-^0 \, c^0 \, c_+^0]^T$ as a linear combination of the right eigenvectors v_1, v_2, v_3 of L, as explained in Appendix A.5:

$$
\begin{bmatrix} c_-^0 \\ c^0 \\ c_+^0 \end{bmatrix} = a_1\, v_1 + a_2\, v_2 + a_3\, v_3
$$

(6.6)

The uniform cubic B-spline subdivision matrix L has eigenvalues $(\lambda_1, \lambda_2, \lambda_3) = (1, 1/2, 1/4)$ and associated right eigenvectors

$$
v_1 = \begin{bmatrix} 1 \\ 1 \\ 1 \end{bmatrix} \quad v_2 = \begin{bmatrix} -1 \\ 0 \\ 1 \end{bmatrix} \quad v_3 = \begin{bmatrix} 2 \\ -1 \\ 2 \end{bmatrix}
$$

The advantage of the representation given in equation (6.6) is that it is very easy to multiply by $(L)^j$:

$$
\begin{aligned}
(L)^j \begin{bmatrix} c_-^0 \\ c^0 \\ c_+^0 \end{bmatrix} &= (L)^j \left(a_1\, v_1 + a_2\, v_2 + a_3\, v_3 \right) \\
&= a_1\, (L)^j\, v_1 + a_2\, (L)^j\, v_2 + a_3\, (L)^j\, v_3 \\
&= a_1\, (\lambda_1)^j\, v_1 + a_2\, (\lambda_2)^j\, v_2 + a_3\, (\lambda_3)^j\, v_3
\end{aligned}
$$

Because $\lambda_1 = 1$ and the other eigenvalues are less than 1, as j approaches infinity, only the first term survives:

$$
\begin{aligned}
\begin{bmatrix} c_-^\infty \\ c^\infty \\ c_+^\infty \end{bmatrix} &= \lim_{j \to \infty} (L)^j \begin{bmatrix} c_-^0 \\ c^0 \\ c_+^0 \end{bmatrix} \\
&= \lim_{j \to \infty} \left(a_1\, (\lambda_1)^j\, v_1 + a_2\, (\lambda_2)^j\, v_2 + a_3\, (\lambda_3)^j\, v_3 \right) \\
&= a_1\, v_1
\end{aligned}
$$

Thus, the limit position c^∞ is the middle entry of $a_1\, v_1$. Since the middle entry of v_1 is 1, the limit position c^∞ is simply a_1. According to Appendix A.5, the coefficient a_1 is computed using the *dominant left eigenvector*, that is, the left eigenvector u_1 associated with the largest eigenvalue. For uniform cubic B-spline subdivision, this eigenvector is $u_1 = \frac{1}{6}[1 \ \ 4 \ \ 1]$. The limit position c^∞ is therefore

$$c^\infty = \boldsymbol{u}_1 \begin{bmatrix} c_-^0 \\ c^0 \\ c_+^0 \end{bmatrix}$$

$$= \frac{1}{6} [1 \ 4 \ 1] \begin{bmatrix} c_-^0 \\ c^0 \\ c_+^0 \end{bmatrix}$$

$$= \frac{c_-^0 + 4\, c^0 + c_+^0}{6}$$

To summarize, a limit position can be computed by multiplying the initial vertex positions by the dominant left eigenvector. For Chaikin's algorithm, the dominant left eigenvector is $\frac{1}{2}[1 \ 1]$, and therefore

$$c^\infty = \frac{c_-^0 + c^0}{2}$$

In general, then, the dominant left eigenvector serves as an evaluation mask since its entries describe a weighted average that can be used to compute a value of the function. Intuitively speaking, evaluation masks carry vertices to their limit positions. Derivative masks can be obtained by considering left eigenvectors associated with eigenvalues less than 1 (Halstead et al. [54] and Hoppe et al. [60] discuss derivative masks in detail).

Although the discussion above described how to send a particular vertex at level 0 to its limiting value, the result can actually be applied to any vertex at any level. An efficient way to render functions, curves, and surfaces defined through subdivision is to apply the subdivision process a few times (typically two or three) to create a reasonably large number of vertices, and then push each vertex to its limit using the evaluation mask.

6.4   Nested spaces and refinable scaling functions

We mentioned in Section 2.2 that a sequence of nested spaces $V^0 \subset V^1 \subset \cdots$ plays a fundamental role in the development of wavelets. In this section we show that subdivision naturally leads to such a sequence of spaces. This fact will be used in subsequent chapters to hierarchically decompose curves and surfaces generated through subdivision.

We begin by recalling that in Section 2.2 we constructed spaces V^j by considering all linear combinations of translated and scaled box functions. Specifically, we need

$$V^j = \text{span}\left\{\phi_0^j(x), \phi_1^j(x), \ldots, \phi_{2^j-1}^j(x)\right\}$$

Stated another way, the functions $\phi_0^j(x), \ldots, \phi_{2^j-1}^j(x)$ form a basis for the space V^j of piecewise-constant functions. In general, basis functions for such nested spaces are called *scaling functions*. The nesting of these spaces is equivalent to the fact that the box scaling functions are *refinable*; that is, each box function at resolution $j-1$ can be written as a linear combination of box functions at resolution j. Referring to Figure 6.7, it is clear that the specific linear combination to use for box functions is

$$\phi_i^{j-1}(x) = 1 \cdot \phi_{2i}^j(x) + 1 \cdot \phi_{2i+1}^j(x)$$

We will now show that every subdivision scheme gives rise to refinable scaling functions and, hence, to nested function spaces. The first step is to note that the splitting and averaging steps of subdivision are both linear in the initial values c_i^0, meaning that the limit function can be written as

$$f(x) = \sum_i c_i^0 \, \phi_i^0(x)$$
(6.7)

for some as yet undetermined functions $\phi_i^0(x)$. These functions will in fact turn out to be the scaling functions. Since the limit curve $f(x)$ defined by the values c_i^0 is the same as the limit curve defined by the values c_i^j for any $j > 0$, we can more generally write

$$f(x) = \cdots = \sum_i c_i^{j-1} \, \phi_i^{j-1}(x) = \sum_i c_i^j \, \phi_i^j(x) = \cdots$$
(6.8)

Matrix notation will allow us to proceed more easily, so we'll let $\mathbf{\Phi}^j(x)$ be a row matrix containing all the scaling functions with superscript j; that is,

$$\mathbf{\Phi}^j(x) = [\phi_0^j(x) \quad \phi_1^j(x) \quad \cdots \,]$$

Now equation (6.8) can be written

$$f(x) = \cdots = \mathbf{\Phi}^{j-1}(x) \, \mathbf{c}^{j-1} = \mathbf{\Phi}^j(x) \, \mathbf{c}^j = \cdots$$
(6.9)

All we have argued so far is that the scaling functions $\phi_i^j(x)$ exist and are somehow determined by the specific subdivision process being used. Next, we will define a space V^j for each j that includes all linear combinations of scaling functions with superscript j:

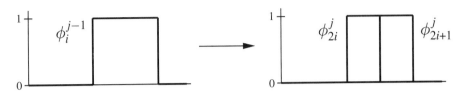

FIGURE 6.7   Box function refinement.

$$V^j := \text{span}\left\{\phi_0^j(x), \phi_1^j(x), \ldots\right\} \tag{6.10}$$

We now show that the scaling functions are refinable, and hence the spaces V^j form a nested sequence—exactly what we will need in Chapter 7 to consider hierarchical decompositions. Because the splitting and averaging steps of subdivision are linear in the values c_i^{j-1}, there is a nonsquare matrix P^j that represents the combined effect of splitting and averaging, and is such that

$$c^j = P^j c^{j-1} \tag{6.11}$$

(The matrix P^j could also be written as a product of the averaging matrix R^j and an additional matrix governing splitting, which we haven't had reason to define explicitly.) For obvious reasons, the matrix P^j is called a *subdivision matrix*. It plays a crucial role in the theory of multiresolution analysis as described in the next chapter.

Next, we observe that according to equation (6.9), we can rewrite a function at subdivision level $j - 1$ as a function at subdivision level j:

$$\mathbf{\Phi}^{j-1}(x)\, c^{j-1} = \mathbf{\Phi}^j(x)\, c^j$$

Substituting equation (6.11) into the equation above gives

$$\mathbf{\Phi}^{j-1}(x)\, c^{j-1} = \mathbf{\Phi}^j(x)\, P^j\, c^{j-1}$$

Because this equation must hold for any values of c^{j-1}, we then have

$$\mathbf{\Phi}^{j-1}(x) = \mathbf{\Phi}^j(x)\, P^j \tag{6.12}$$

This last equation is known as a *refinement relation* for the scaling functions. It states that each of the coarse scaling functions $\phi_i^{j-1}(x)$ can be written as a linear combination of the fine scaling functions $\phi_i^j(x)$.

Remark: There is a straightforward recipe for computing a scaling function $\phi_k^j(x)$, at least approximately: simply run the subdivision procedure starting with a sequence of values $c_i^j = \delta_{i,k}$ at level j. The limit function $f(x)$ is exactly $\phi_k^j(x)$ since, by equation (6.8), $f(x)$ is given by just one term $c_k^j \phi_k^j(x)$ and, by construction, $c_k^j = 1$. In Figure 6.8, this method has been used to plot several different endpoint-interpolating B-spline scaling functions using the averaging matrices from Section 6.2. ∎

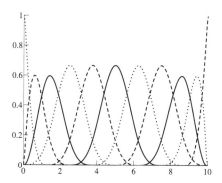

FIGURE 6.8 Endpoint-interpolating cubic B-spline scaling functions generated by repeated subdivision.

Remark: In the case of uniform subdivision, the functions $\phi_i^0(x)$ all have the same shape; they are just shifted relative to one another. In fact, the scaling functions are always shifted by an integer distance. Thus, there is a single function $\phi(x)$, sometimes called the *father scaling function*, from which all others are determined according to $\phi_i^0(x) = \phi(x - i)$. The fact that uniform subdivision produces shifted copies of a single father function is demonstrated in Figure 6.9 for the Daubechies subdivision scheme introduced in Section 6.1.

The scaling functions $\phi_i^1(x)$ at resolution level 1 are also shifted versions of one another, but they are only shifted by half-integers. Moreover, since in the uniform stationary case the same subdivision procedure is used in each iteration, the functions $\phi_i^1(x)$ have the same basic shape as $\phi(x)$, but they are shrunk horizontally by a factor of 2. In general, $\phi_i^j(x) = \phi(2^j x - i)$. This change in width by factors of two is often referred to as *dilation*. ∎

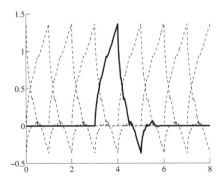

FIGURE 6.9   Uniform subdivision produces scaling functions that are shifted copies of one another.

Remark: For uniform subdivision, the refinement relation for the father scaling function also simplifies considerably to

$$\phi(x) = \sum_i p_i\, \phi(2x - i)$$

Thus, the father scaling function has the rather remarkable property that it can be represented as a linear combination of dilates and translates of itself. The *refinement coefficients* p_i for various uniform schemes are listed in Table 6.1. ∎

TABLE 6.1   Refinement coefficients for scaling functions resulting from uniform subdivision schemes.

Scheme	Refinement Coefficients
Piecewise-constant	$(1, 1)$
Piecewise-linear	$\frac{1}{2}(1, 2, 1)$
Piecewise-quadratic	$\frac{1}{4}(1, 3, 3, 1)$
Piecewise-cubic	$\frac{1}{8}(1, 4, 6, 4, 1)$
Daubechies D_4	$\frac{1}{4}(1 + \sqrt{3}, 3 + \sqrt{3}, 3 - \sqrt{3}, 1 - \sqrt{3})$
DLG	$\frac{1}{16}(-1, 0, 9, 16, 9, 0, -1)$

In summary, we have shown that the scaling functions are refinable and that P^j serves as both the subdivision matrix for the initial values c^j and the refinement matrix for the scaling functions $\Phi^j(x)$. As a result, we know that any collection of functions that is refinable can be generated through subdivision, with the subdivision matrix and the refinement matrix being one and the same.

THE THEORY OF
MULTIRESOLUTION ANALYSIS

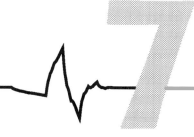

1. Multiresolution analysis — 2. Orthogonal wavelets — 3. Semiorthogonal wavelets — 4. Biorthogonal wavelets — 5. Summary

In the previous chapter, we showed that only scaling functions defined through subdivision are refinable. This result has the important consequence that the *only* functions that can be used to form nested linear function spaces are those defined through subdivision.

Moreover, as you may recall from Chapter 2, the starting point for multiresolution analysis is a nested set of linear spaces

$$V^0 \subset V^1 \subset V^2 \subset \cdots$$

Thus, the subdivision process is also fundamentally tied to multiresolution analysis itself.

In this chapter, we'll develop the mathematical framework of shift-variant multiresolution analysis in detail. We will then discuss three different classes of wavelet bases—orthogonal, semiorthogonal, and biorthogonal—and give examples of each. The different types of wavelet bases allow us to trade off, to varying degrees, such desirable properties as orthogonality, compact support, degrees of smoothness, and symmetry.

7.1 Multiresolution analysis

As mentioned above, the starting point for multiresolution analysis is a nested set of linear function spaces $V^0 \subset V^1 \subset \cdots$, with the resolution of functions in V^j increasing with j. As mentioned in Section 6.4, the basis functions for the space V^j are called scaling functions.

The next step in multiresolution analysis is to define the *wavelet spaces*, denoted by W^j. In order to encompass a wider variety of wavelet constructions than we have until now, we'll define the wavelet spaces in a looser way here than we did in Chapter 2. Here, we will simply require each wavelet space W^j to be the *complement* of V^j in V^{j+1}. Thus, any function in V^{j+1} can be written as the sum of a unique function in V^j and a unique function in W^j. The functions we choose as a basis for W^j are called *wavelets*. Note that in contrast to the definition given in Chapter 2, we're no longer requiring the wavelets in W^j to necessarily be orthogonal to the scaling functions in V^j.

The rest of our discussion of multiresolution analysis will focus on wavelets defined on a bounded domain for curves and (in Chapter 10) for compact surfaces of arbitrary topological type. At various points in the development, we will also discuss the more traditional setting of wavelets on the unbounded real line. In the bounded case, each space V^j has a finite basis, allowing us to use matrix notation in much of what follows, as did Lounsbery et al. [75] and Quak and Weyrich [98]. Although all the examples in this section are for one-dimensional wavelets, the theoretical development also holds for the surface wavelets described in Chapter 10.

7.1.1 Refinement

As we mentioned in Section 6.4, it is often convenient to put the different scaling functions $\phi_i^j(x)$ for a given level j together into a single row matrix. We'll denote the dimension of V^j by $v(j)$, and write

$$\mathbf{\Phi}^j(x) := [\phi_0^j(x) \quad \cdots \quad \phi_{v(j)-1}^j(x)]$$

Similarly, if $w(j)$ is the dimension of W^j, then we can write the wavelets at level j as

$$\mathbf{\Psi}^j(x) := [\psi_0^j(x) \quad \cdots \quad \psi_{w(j)-1}^j(x)]$$

Because W^j is the complement of V^j in V^{j+1}, the dimensions of these spaces satisfy $v(j+1) = v(j) + w(j)$.

In Section 6.4, we established that having nested subspaces V^j is equivalent to having scaling functions that are refinable. That is, for all $j = 1, 2, \ldots$ there must exist a matrix of constants $\boldsymbol{P}^j$ such that

$$\boldsymbol{\Phi}^{j-1}(x) = \boldsymbol{\Phi}^j(x)\, \boldsymbol{P}^j \tag{7.1}$$

Note that since V^j and V^{j-1} have dimensions $v(j)$ and $v(j-1)$, respectively, $\boldsymbol{P}^j$ is a $v(j) \times v(j-1)$ matrix.

Since the wavelet space W^{j-1} is by definition also a subspace of V^j, we can write the wavelets $\boldsymbol{\Psi}^{j-1}(x)$ as linear combinations of the scaling functions $\boldsymbol{\Phi}^j(x)$. Therefore, there is a $v(j) \times w(j-1)$ matrix of constants $\boldsymbol{Q}^j$ satisfying

$$\boldsymbol{\Psi}^{j-1}(x) = \boldsymbol{\Phi}^j(x)\, \boldsymbol{Q}^j \tag{7.2}$$

Example: In the Haar basis, at a particular level j there are $v(j) = 2^j$ scaling functions and $w(j) = 2^j$ wavelets. Thus, there must be refinement matrices describing how the two scaling functions in V^1 and the two wavelets in W^1 can be made from the four scaling functions in V^2 (refer back to Figures 2.2 and 2.3):

$$\boldsymbol{P}^2 = \begin{bmatrix} 1 & 0 \\ 1 & 0 \\ 0 & 1 \\ 0 & 1 \end{bmatrix} \quad \text{and} \quad \boldsymbol{Q}^2 = \begin{bmatrix} 1 & 0 \\ -1 & 0 \\ 0 & 1 \\ 0 & -1 \end{bmatrix}$$

∎

Remark: In the case of wavelets constructed on the unbounded real line, the wavelets $\psi_i^j(x)$ all have the same shape; in fact, they are all just shifted and scaled versions of a single function $\psi(x)$, called the *mother wavelet*. In general, $\psi_i^j(x) = \psi(2^j x - i)$.

In addition, in this shift-invariant case, the columns of $\boldsymbol{P}^j$ are shifted versions of one another, as are the columns of $\boldsymbol{Q}^j$. One column therefore characterizes each matrix, so $\boldsymbol{P}^j$ and $\boldsymbol{Q}^j$ are completely determined by sequences $(\ldots, p_{-1}, p_0, p_1, \ldots)$ and $(\ldots, q_{-1}, q_0, q_1, \ldots)$, which also do not depend on j. Thus, equations (7.1) and (7.2) often appear in the literature as expressions of the form

$$\phi(x) = \sum_i p_i\, \phi(2x - i)$$

$$\psi(x) = \sum_i q_i\, \phi(2x - i)$$

∎

Note that equations (7.1) and (7.2) can also be expressed as a single equation, using block-matrix notation:

$$\left[\Phi^{j-1} \mid \Psi^{j-1}\right] = \Phi^j \left[P^j \mid Q^j\right] \tag{7.3}$$

This equation is referred to as a *two-scale relation* for scaling functions and wavelets.

Example: Substituting the matrices from the previous example into equation (7.3) along with the appropriate basis functions yields

$$[\phi_0^1 \ \phi_1^1 \ \psi_0^1 \ \psi_1^1] = [\phi_0^2 \ \phi_1^2 \ \phi_2^2 \ \phi_3^2] \begin{bmatrix} 1 & 0 & 1 & 0 \\ 1 & 0 & -1 & 0 \\ 0 & 1 & 0 & 1 \\ 0 & 1 & 0 & -1 \end{bmatrix} \qquad \blacksquare$$

7.1.2 The filter bank

Now that we've described how subdivision and matrices relate to scaling functions and wavelets, we need to show how matrix notation can also be used to perform a wavelet transform. The material here generalizes the hierarchical decomposition of a function that was outlined in Section 2.1 for the Haar basis.

Consider a function in some approximation space V^j. Let's assume we have the coefficients of this function in terms of some scaling function basis. We can write these coefficients as a column matrix of values $c^j = [c_0^j \ \cdots \ c_{v(j)-1}^j]^T$. The coefficients c_i^j could, for example, be thought of as pixel colors, or alternatively, as the x- or y-coordinates of a curve's control points in $\mathbb{R}^2$.

Suppose we wish to create a low-resolution version c^{j-1} of c^j with a smaller number of coefficients $v(j-1)$. The standard approach for creating the $v(j-1)$ values of c^{j-1} is to use some form of linear filtering and down-sampling on the $v(j)$ entries of c^j. This process can be expressed as a matrix equation

$$c^{j-1} = A^j c^j \tag{7.4}$$

where A^j is a constant $v(j-1) \times v(j)$ matrix.

Since c^{j-1} contains fewer entries than c^j, it is intuitively clear that some amount of detail is lost in this filtering process. If A^j is appropriately chosen, it is possible to capture the lost detail as another column matrix d^{j-1}, computed by

$$d^{j-1} = B^j c^j \qquad (7.5)$$

where B^j is a constant $w(j-1) \times v(j)$ matrix related to A^j. The pair of matrices A^j and B^j are called *analysis filters*. The process of splitting the coefficients c^j into a low-resolution version c^{j-1} and detail d^{j-1} is called *analysis* or *decomposition*.

If A^j and B^j are chosen appropriately, then the original coefficients c^j can be recovered from c^{j-1} and d^{j-1} by using the matrices P^j and Q^j from the previous section:

$$c^j = P^j c^{j-1} + Q^j d^{j-1} \qquad (7.6)$$

Recovering c^j from c^{j-1} and d^{j-1} is called *synthesis* or *reconstruction*, and, in this context, P^j and Q^j are called *synthesis filters*.

Intuitively, the first term on the right side of equation (7.6) indicates that the initial setting of the coefficients c^j is obtained by "interpolating up" the coarse coefficients c^{j-1} to the finer scale. The scaling function coefficients are interpolated up by executing one step of subdivision—that is, by multiplying by P^j, or, equivalently, by performing splitting and averaging. The second term can be thought of as a perturbation to the first term and is obtained by interpolating up the wavelet coefficients. Wavelet coefficients are interpolated up by multiplying by Q^j.

Example: In the unnormalized Haar basis, the matrices A^2 and B^2 are given by:

$$A^2 = \frac{1}{2}\begin{bmatrix} 1 & 1 & 0 & 0 \\ 0 & 0 & 1 & 1 \end{bmatrix}$$

$$B^2 = \frac{1}{2}\begin{bmatrix} 1 & -1 & 0 & 0 \\ 0 & 0 & 1 & -1 \end{bmatrix}$$

These matrices represent the averaging and differencing operations described in Section 2.1. ∎

Remark: Once again, the matrices for wavelets constructed on the unbounded real line have a simple structure: the rows of A^j are shifted versions of each other, as are the rows of B^j. Thus, the analysis equations (7.4) and (7.5) often appear in the literature as

$$c_k^{j-1} = \sum_l a_{l-2k} c_l^j \qquad (7.7)$$

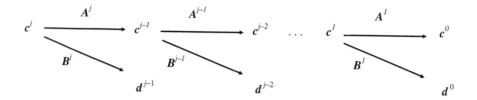

FIGURE 7.1   The filter bank.

$$d_k^{j-1} = \sum_1 b_{1-2k}\, c_1^j$$

(7.8)

where the sequences $(\ldots, a_{-1}, a_0, a_1, \ldots)$ and $(\ldots, b_{-1}, b_0, b_1, \ldots)$ are the entries in a row of A^j and B^j, respectively. Similarly, equation (7.6) for reconstruction often appears as

$$c_k^j = \sum_1 \left(p_{k-21} c_1^{j-1} + q_{k-21} d_1^{j-1} \right)$$

∎

Note that the procedure for splitting c^j into a low-resolution part c^{j-1} and a detail part d^{j-1} can be applied recursively to the low-resolution part c^{j-1}. Thus, the original coefficients can be expressed as a hierarchy of lower-resolution parts $c^0, \ldots, c^{j-1}$ and details $d^0, \ldots, d^{j-1}$, as shown in Figure 7.1. This recursive process is known as a *filter bank*.

Since the original coefficients c^j can be recovered from the sequence $c^0, d^0, d^1, \ldots, d^{j-1}$, this sequence can be thought of as a transform of the original coefficients, known as a *wavelet transform*. Note that the total size of the transform $c^0, d^0, d^1, \ldots, d^{j-1}$ is the same as that of the original version c^j, so no extra storage is required. (However, the scaling function and wavelet coefficients may require more bits to retain the accuracy of the original values.)

In general, the analysis filters A^j and B^j are not necessarily transposed multiples of the synthesis filters, as was the case for the Haar basis. Rather, A^j and B^j are formed by the matrices satisfying the relation

$$\left[\boldsymbol{\Phi}^{j-1} \mid \boldsymbol{\Psi}^{j-1} \right] \left[\dfrac{A^j}{B^j} \right] = \boldsymbol{\Phi}^j$$

(7.9)

Note that

$$[\, \boldsymbol{P}^j \mid \boldsymbol{Q}^j \,]$$

and

$$\left[\frac{\boldsymbol{A}^j}{\boldsymbol{B}^j} \right]$$

are both square matrices. Thus, combining equations (7.3) and (7.9) gives

$$\left[\frac{\boldsymbol{A}^j}{\boldsymbol{B}^j} \right] = [\, \boldsymbol{P}^j \mid \boldsymbol{Q}^j \,]^{-1} \tag{7.10}$$

Although we have not been specific about how to choose matrices $\boldsymbol{A}^j$, $\boldsymbol{B}^j$, $\boldsymbol{P}^j$, and $\boldsymbol{Q}^j$, it should be clear from equation (7.10) that

$$\left[\frac{\boldsymbol{A}^j}{\boldsymbol{B}^j} \right]$$

and

$$[\, \boldsymbol{P}^j \mid \boldsymbol{Q}^j \,]$$

must at least be invertible. In the remainder of this chapter, we will explore three different constructions of wavelets that differ primarily in the way they develop the wavelet refinement matrices $\boldsymbol{Q}^j$.

7.2 Orthogonal wavelets

An *orthogonal basis* is one in which each basis function is orthogonal to every other basis function. An *orthogonal multiresolution basis* is one in which the scaling functions are orthogonal to one another, the wavelets are orthogonal to one another, and each of the wavelets is orthogonal to every coarser scaling function. We call wavelets that satisfy these conditions *orthogonal wavelets*.

We pointed out in Chapter 2 that the Haar basis is orthogonal, and we relied on that fact in developing a compression algorithm: the L^2 error of an approximation was easily computed from the magnitudes of the coefficients that had been set to zero. This simple relationship between coefficient size and error exists only when we use a basis that is orthogonal.

In mathematical notation, the conditions defining an orthogonal wavelet basis can be written as

$$\left. \begin{array}{l} \langle \, \phi_k^j \mid \psi_\ell^j \, \rangle = \delta_{k,\ell} \\ \langle \, \psi_k^j \mid \psi_\ell^j \, \rangle = \delta_{k,\ell} \\ \langle \, \phi_k^j \mid \psi_\ell^j \, \rangle = 0 \end{array} \right\} \text{ for all } j, k, \text{ and } \ell \qquad (7.11)$$

Actually, these conditions define an *orthonormal* wavelet basis, since the inner product of a basis function with itself is constrained to be 1. An orthogonal basis need not have this normalization, but in the rest of this development we will assume without loss of generality that an orthogonal wavelet basis is also orthonormal.

7.2.1   Implications of orthogonality

To deal with all the inner products in equation (7.11) simultaneously, let's define some new notation for a matrix of inner products. Let $f(x) = [f_0(x) \ \ f_1(x) \ \ \cdots \ ]$ and $g(x) = [g_0(x) \ \ g_1(x) \ \ \cdots \ ]$ denote two row matrices of functions. We will denote by $[\langle f \mid g \rangle]$ the matrix whose (k, ℓ) entry is $\langle f_k \mid g_\ell \rangle$. Unlike an inner product between a pair of functions, $[\langle f \mid g \rangle]$ is not necessarily symmetric, but the following relations between f, g, and a constant matrix M do hold:

$$[\langle f \mid g \rangle] = [\langle g \mid f \rangle]^{\mathrm{T}}$$

$$[\langle f \mid g \, M \rangle] = [\langle f \mid g \rangle] \, M \qquad (7.12)$$

$$[\langle f M \mid g \rangle] = M^{\mathrm{T}} [\langle f \mid g \rangle]$$

We will also use I to denote the identity matrix and 0 to denote the zero matrix or zero vector. Armed with this notation, we can rewrite the orthogonality conditions in equation (7.11) as

$$[\langle \, \Phi^j \mid \Phi^j \, \rangle] = I$$

$$[\langle \, \Psi^j \mid \Psi^j \, \rangle] = I$$

$$[\langle \, \Phi^j \mid \Psi^j \, \rangle] = 0$$

An even more concise statement of the orthogonality conditions results if we group all the basis functions together:

$$
\left[\left\langle \ [\, \boldsymbol{\Phi}^j |\ \boldsymbol{\Psi}^j\,]\ \middle|\ [\, \boldsymbol{\Phi}^j |\ \boldsymbol{\Psi}^j\,]\ \right\rangle \right] = \left[\begin{array}{c|c} [\langle\, \boldsymbol{\Phi}^j\, |\, \boldsymbol{\Phi}^j\, \rangle] & [\langle\, \boldsymbol{\Phi}^j\, |\, \boldsymbol{\Psi}^j\, \rangle] \\ \hline [\langle\, \boldsymbol{\Psi}^j\, |\, \boldsymbol{\Phi}^j\, \rangle] & [\langle\, \boldsymbol{\Psi}^j\, |\, \boldsymbol{\Psi}^j\, \rangle] \end{array} \right] = I
$$

In order to show how orthogonality affects the P^j and Q^j matrices, let's change the superscript from j to $j-1$ and rewrite the orthogonality conditions above as

$$
\left[\left\langle\ [\, \boldsymbol{\Phi}^{j-1} |\ \boldsymbol{\Psi}^{j-1}\,]\ \middle|\ [\, \boldsymbol{\Phi}^{j-1} |\ \boldsymbol{\Psi}^{j-1}\,]\ \right\rangle \right] = I
$$

Now, if we substitute equation (7.3) into the previous equation in two places, we get

$$
\left[\left\langle\ \boldsymbol{\Phi}^j\, [\, P^j |\ Q^j\,]\ \middle|\ \boldsymbol{\Phi}^j\, [\, P^j |\ Q^j\,]\ \right\rangle \right] = I
$$

We can use the properties of inner product matrices listed in equation (7.12) to get

$$
[\, P^j\, |\ Q^j\,]^\mathrm{T}\, [\langle\, \boldsymbol{\Phi}^j\, |\ \boldsymbol{\Phi}^j\, \rangle]\, [\, P^j\, |\ Q^j\,] = I
$$

Finally, we notice that $[\langle\, \boldsymbol{\Phi}^j\, |\ \boldsymbol{\Phi}^j\, \rangle] = I$ and arrive at the result

$$
[\, P^j\, |\ Q^j\,]^\mathrm{T} = [\, P^j\, |\ Q^j\,]^{-1}
$$

Thus, for orthogonal wavelets, the inverse of the combined scaling function and wavelet synthesis matrix

$$
[\, P^j\, |\ Q^j\,]
$$

is the same as its transpose; a matrix with this property is called an *orthogonal matrix*. This fact, in combination with equation (7.10), indicates that

$$A^j = (P^j)^T \quad \text{and} \quad B^j = (Q^j)^T$$

7.2.2 Daubechies wavelets

At this point, we know that subdivision schemes can be used to define scaling functions and nested function spaces. It seems natural to ask whether we can use subdivision to develop an orthogonal wavelet basis that is somehow better than the Haar basis—in particular, finding a set of smooth basis functions would be nice. This goal was achieved by Daubechies [24] in her development of orthonormal, compactly supported wavelets for the infinite real line. Since in this book we are focusing on constructions of wavelets on bounded intervals, we won't present a derivation of the Daubechies wavelets here. Instead we'll simply illustrate the results of her work with an example of the synthesis and analysis filters.

The D_4 construction, so named because its filters each have four entries, makes use of the following sequences:

$$p = a = \frac{1}{4\sqrt{2}} (1 + \sqrt{3}, \ 3 + \sqrt{3}, \ 3 - \sqrt{3}, \ 1 - \sqrt{3})$$

$$q = b = \frac{1}{4\sqrt{2}} (1 - \sqrt{3}, \ -3 + \sqrt{3}, \ 3 + \sqrt{3}, \ -1 - \sqrt{3})$$

The p sequence represents the nonzero entries of each column of P^j, and the q sequence gives the entries in the columns of Q^j. Likewise, the a and b sequences represent the rows of A^j and B^j, respectively. Note that the wavelet filter sequence can be generated by reversing the order of the entries in the scaling function sequence and alternating their signs. This recipe for creating a wavelet sequence from a scaling function sequence is common to many wavelet constructions on the infinite real line; such sequences are known as *quadrature mirror filters*.

The D_4 scaling function and wavelet are shown in Figure 7.2. Both of the functions are asymmetric and nowhere differentiable, although they are continuous. These basis functions are members of a whole family of bases developed by Daubechies, including bases whose smoothness increases with the supports of the basis functions.

We have already discussed in Section 6.4 how we can graph a scaling function: we start with a sequence of coefficients $c_i^j = \delta_{i,k}$ at level j, subdivide a number of times, and graph the results. We can subdivide either by splitting and averaging, as described in the previous chapter, or by inserting zeros between all the coefficients in the current sequence and convolving with the refinement sequence p. (Inserting zeros before convolving is analogous to splitting before averaging.) The method for graphing a wavelet is very similar. To get the k-th wavelet at level j, we start with a sequence of wavelet coefficients $d_i^j = \delta_{i,k}$, insert zeros between the

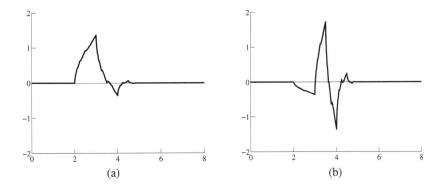

FIGURE 7.2   Daubechies basis functions: (a) the D_4 scaling function; (b) the D_4 wavelet.

elements of this sequence, and convolve with q to create a sequence of scaling function coefficients c^{j+1}. From there, we can repeatedly insert zeros and convolve with the sequence p to reach the wavelet's limit function.

> **Remark:**   The p and q sequences given above are the correct ones to use for the reconstruction of a function from its scaling function and wavelet coefficients. However, in order to make the graphs in Figure 7.2 using recursive subdivision, we have to use slightly different sequences. The reason for this inconsistency is that a subdivision scheme cannot converge unless the entries in each row of $\boldsymbol{P}^j$ sum to 1. For a uniform stationary scheme (that is, a scheme that uses a single sequence p everywhere), this condition is equivalent to requiring that the odd entries of p sum to 1 and that the even entries of p sum to 1. It turns out that if we multiply the p sequence given above by $\sqrt{2}$, we have a convergent subdivision scheme. To be consistent, we also multiply the q sequence by $\sqrt{2}$ before applying subdivision to graph a wavelet. ∎

7.3   Semiorthogonal wavelets

Orthogonality is not the only desirable property in a wavelet basis. Here is a list of some other properties that may be more or less important when constructing a wavelet basis for a particular application:

- *Compact support.* The supports of scaling functions and wavelets are related to the spread of the nonzero entries in $\boldsymbol{P}^j$ and $\boldsymbol{Q}^j$. A more compact support improves the efficiency of decomposing and reconstructing functions using a filter-bank algorithm.

- *Smoothness.* Smooth functions are best represented with smooth bases. For example, piecewise-cubic basis functions can more efficiently represent a smooth curve than can piecewise-linear functions. However, greater smoothness often comes at the expense of wider supports.

- *Symmetry.* For some applications, it may be important to use scaling functions and wavelets that are symmetric (or antisymmetric) about their centers.

- *Vanishing moments.* A wavelet $\psi(x)$ is said to have n *vanishing moments* if the integral $\int \psi(x)\, x^k\, dx$ is identically zero for $k = 0, \ldots, n - 1$ but not for $k = n$. Wavelets with more vanishing moments are often desirable for applications that require numerical approximations of smooth operators.

Limiting ourselves to orthogonal constructions of wavelets can be overly restrictive. In fact, there are no wavelet bases (besides the Haar basis) that consist of functions that are at once orthogonal, compactly supported, *and* symmetric [25]. Thus, if we desire smooth symmetric wavelets with compact supports, we must be willing to sacrifice orthogonality. However, the loss is not always severe: we can often construct a multiresolution basis in which the wavelets of a given resolution are at least orthogonal to all coarser scaling functions, though not necessarily orthogonal to each other. Wavelets designed to meet this criterion are referred to as *semiorthogonal wavelets*, or *pre-wavelets*. In mathematical terms, a semiorthogonal wavelet basis satisfies the conditions

$$\langle\, \phi_k^j \mid \psi_\ell^j\, \rangle = 0 \quad \text{for all } j, k, \text{ and } \ell \tag{7.13}$$

Note that orthogonal wavelets are a special case of semiorthogonal wavelets, in which both the scaling functions are orthogonal to one another and the wavelets are orthogonal to one another.

7.3.1 Implications of semiorthogonality

To see what semiorthogonality implies about the matrices $\boldsymbol{P}^j$ and $\boldsymbol{Q}^j$, we'll use the notation we introduced in Section 7.2.1 for a matrix of inner products. First, we rewrite equation (7.13) as

$$[\langle\, \boldsymbol{\Phi}^j \mid \boldsymbol{\Psi}^j\, \rangle] = \boldsymbol{0}$$

If we change the superscript j to $j - 1$, we can substitute the refinement relations for $\boldsymbol{P}^j$ and $\boldsymbol{Q}^j$ in equations (7.1) and (7.2) into the previous equation, to get

$$(P^j)^{\mathrm{T}} [\langle \, \Phi^j \mid \Phi^j \, \rangle] \, Q^j = 0 \qquad\qquad (7.14)$$

Given a set of scaling functions Φ^j and a refinement matrix P^j, we would like to find a (good) wavelet refinement matrix Q^j, which will also determine the wavelets Ψ^j. We must therefore solve the system of equations in (7.14) for the matrix Q^j.

A matrix equation with a right-hand side of zero, such as equation (7.14), is known as a *homogeneous system* of equations. For the sake of simplifying the current discussion, let's define $M^j := (P^j)^{\mathrm{T}} [\langle \, \Phi^j \mid \Phi^j \, \rangle]$. The set of all possible solutions to equation (7.14) is called the *null space* of M^j. A null space is nothing more than a finite-dimensional vector space—in this case, the space of column vectors that yield zero when they are multiplied by any row of M^j. We are interested in finding columns of Q^j that form a basis for this space.

There are a multitude of bases for the null space of a rectangular matrix like M^j, implying that there are many different wavelet bases for a given wavelet space W^j. Assuming we already know M^j and we want to choose Q^j to define the wavelets, we need to impose further constraints in addition to the orthogonality requirement in equation (7.14). If we want our wavelets to have compact supports, we can force the columns of Q^j to have the fewest possible consecutive nonzero entries. If we want our wavelets to be symmetric about their centers, we can enforce symmetry in the entries in each column of Q^j.

7.3.2 Spline wavelets

As an example of a semiorthogonal basis, we now turn to a class of wavelets that is particularly useful for representing multiresolution curves: *spline wavelets*. These wavelets, which are built from B-splines, have been developed to a large extent by Chui and colleagues [17, 18]. Of particular interest for curve-editing applications are endpoint-interpolating B-splines. In what follows, we briefly sketch the ideas behind the construction of endpoint-interpolating B-spline wavelets, following the technique used by Finkelstein and Salesin [37] to make wavelets for the cubic case.

Note that the results of the derivation are summarized in Appendix B, where we give the synthesis filters for the piecewise-constant (Haar), linear, quadratic, and cubic cases. The Matlab code that was used to generate those filters appears in Appendix C.

B-spline scaling functions

Our first step is to define the scaling functions for a nested set of function spaces. For this construction, we'll use the endpoint-interpolating nonuniform B-splines [35] defined on the unit interval. Once we choose the degree m of the basis functions we want, we can define

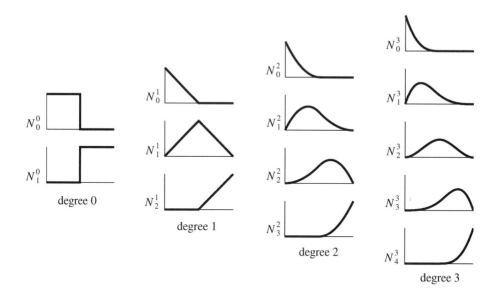

FIGURE 7.3   Endpoint-interpolating B-spline scaling functions for $V^1(m)$ with degree $m = 0, 1, 2,$ and 3.

$V^j(m)$ to be the set of $2^j + m$ nonuniform B-splines that are constructed from the knot sequence

$$(x_0, \ldots, x_{2^j+2m}) = \frac{1}{2^j} (\underbrace{0, \ldots, 0}_{m+1 \text{ times}}, 1, 2, \ldots, 2^j - 2, 2^j - 1, \underbrace{2^j, \ldots, 2^j}_{m+1 \text{ times}})$$

Examples of the spline scaling functions at level $j = 1$ are shown in Figure 7.3 for various degrees m.

It is not difficult to show that the spaces $V^0(m)$, $V^1(m)$, ... are nested as required by multiresolution analysis. The theory of knot insertion for B-splines can be used to develop expressions for the entries of the refinement matrix P^j (see Chui and Quak [18] for details). The columns of P^j are sparse, reflecting the fact that the B-spline basis functions are locally supported. The first and last m columns of P^j are relatively complicated, but the remaining (interior) columns are shifted versions of column $m + 1$. Moreover, the entries of these interior columns are, up to a common factor of 2^{-m}, given by binomial coefficients.

Example: In the case of cubic splines ($m = 3$), the matrix P^j for $j \geq 3$ has the form

$$
P^j = \frac{1}{8}
\begin{bmatrix}
8 \\
4 & 4 \\
& 6 & 2 \\
& \frac{3}{2} & \frac{11}{2} & 1 \\
& & 4 & 4 \\
& & 1 & 6 & 1 \\
& & & 4 & 4 \\
& & & 1 & 6 \\
& & & & 4 & \cdot \\
& & & & 1 & \cdot & 1 \\
& & & & & \cdot & 4 \\
& & & & & & 6 & 1 \\
& & & & & & 4 & 4 \\
& & & & & & 1 & \frac{11}{2} & \frac{3}{2} \\
& & & & & & & 2 & 6 \\
& & & & & & & & 4 & 4 \\
& & & & & & & & & 8
\end{bmatrix}
$$

where blank entries are taken to be zero and the dots indicate that the previous column is repeated, shifted down by two rows each time. ∎

Inner product

The second step in designing a semiorthogonal basis is the choice of an inner product. We'll simply use the standard inner product here:

$$
\langle f \mid g \rangle := \int_0^1 f(x)\, g(x)\, dx
$$

B-spline wavelets

To complete our development of a semiorthogonal B-spline multiresolution analysis, we need to find basis functions for the spaces W^j that are orthogonal complements to the spaces V^j. It is important to realize that once we have chosen scaling functions and their refinement matrices P^j, the wavelet matrices Q^j are somewhat constrained, though not completely determined. If our wavelets are to be orthogonal to the scaling functions, the columns of Q^j must form a basis for the null space of $M^j := (P^j)^{\mathrm{T}} [\langle \Phi^j \mid \Phi^j \rangle]$. Because the dimension of $V^j(m)$ is

$2^j + m$, we know M^j has $2^j + m$ columns and only $2^{j-1} + m$ rows. The difference between these two numbers tells us that the dimension of the null space of M^j is 2^{j-1}. Thus, there are 2^{j-1} wavelets in $W^j(m)$, each defined by a column of Q^j.

As we already mentioned, there are many bases for the null space of a matrix. To uniquely determine the Q^j matrices, we need to decide what properties the wavelets should have. The approach taken by Finkelstein and Salesin [37] and the one we will take here is to require the columns of Q^j to be sparse and further require a minimal number of consecutive nonzero entries. By putting as many zeros as possible at the top and bottom of each column of Q^j, we guarantee that the wavelets will have compact supports.

The structure of the resulting Q^j matrices is similar to that of the P^j matrices: the first and last m columns are unusual, but the remaining interior columns are all shifted copies of a single sequence.

Example:　To illustrate the structure of the Q^j matrices, here's what the Q^j matrix for endpoint-interpolating cubic B-splines looks like for $j \geq 4$, with blanks indicating zeros, $\times$ indicating the uninteresting noninteger values, and β representing a normalization constant:

$$
Q^j = \beta
\begin{bmatrix}
\times \\
\times & \times \\
\times & \times & \times \\
\times & \times & \times & -1 \\
\times & \times & \times & 124 \\
\times & \times & \times & -1677 & -1 \\
\times & \times & \times & 7904 & 124 \\
\times & \times & \times & -18482 & -1677 \\
\times & \times & 24264 & 7904 \\
\times & \times & -18482 & -18482 & & -1 \\
\times & 7904 & 24264 & & 124 \\
\times & -1677 & -18482 & \cdot & -1677 & \times \\
& 124 & 7904 & \cdot & 7904 & \times \\
& -1 & -1677 & \cdot & -18482 & \times & \times \\
& & 124 & & 24264 & \times & \times \\
& & -1 & & -18482 & \times & \times & \times \\
& & & & 7904 & \times & \times & \times \\
& & & & -1677 & \times & \times & \times \\
& & & & 124 & \times & \times & \times \\
& & & & -1 & \times & \times & \times \\
& & & & & & \times & \times & \times \\
& & & & & & & \times & \times \\
& & & & & & & & \times
\end{bmatrix}
$$

The matrix above and the other examples that appear in Appendix B reveal a simple structure for the locations of the nonzero entries. The values of the entries can be found by considering each column one at a time and solving the linear system of constraints in equation (7.14). The Matlab code in Appendix C computes the Q^j matrices in this way. Figure 7.4 illustrates some typical endpoint-interpolating B-spline wavelets that result from these constructions.

B-spline filter bank

At this point, we have completed the steps in designing a semiorthogonal wavelet basis. However, in order to use spline wavelets we will need to implement a filter-bank procedure incorporating the analysis filters A^j and B^j. These matrices allow us to determine c^{j-1} and d^{j-1} from c^j using matrix multiplication as is done in equations (7.4) and (7.5). As discussed in Section 7.1.2, the analysis filters are uniquely determined by the inverse relation in equation (7.10):

$$\begin{bmatrix} A^j \\ \hline B^j \end{bmatrix} = [\, P^j \mid Q^j \,]^{-1}$$

However, when implementing the filter-bank procedure for spline wavelets, it is generally *not* a good idea to form the filters A^j and B^j explicitly. Although P^j and Q^j are sparse, having only $O(m)$ entries per column, A^j and B^j are in general dense, so that matrix–vector multiplication would require quadratic instead of linear time.

Fortunately, there is a better approach. The idea is to notice that c^{j-1} and d^{j-1} can be computed from c^j by solving the sparse linear system

$$[\, P^j \mid Q^j \,] \begin{bmatrix} c^{j-1} \\ \hline d^{j-1} \end{bmatrix} = c^j$$

In order to solve this system for

$$\begin{bmatrix} c^{j-1} \\ \hline d^{j-1} \end{bmatrix}$$

the matrix

$$[\, P^j \mid Q^j \,]$$

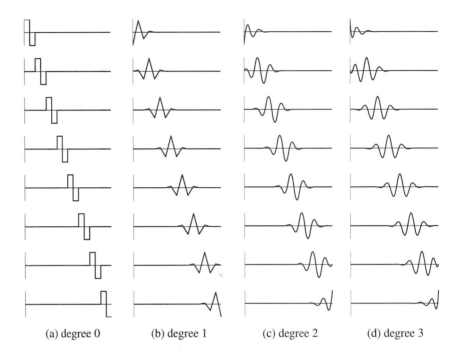

(a) degree 0   (b) degree 1   (c) degree 2   (d) degree 3

FIGURE 7.4   Endpoint-interpolating B-spline wavelets for $W^3(m)$ with degree $m = 0, 1, 2,$ and 3.

can first be made into a banded matrix simply by reordering its columns so that the columns of P^j and Q^j are interspersed. The resulting banded system can then be solved in linear time using LU decomposition [97], and the solution reordered to give the desired answer. Thus the filter-bank operation can be accomplished without ever forming and using A^j and B^j explicitly.

7.3.3   Designing semiorthogonal wavelets

Here's a summary of the steps involved in constructing a semiorthogonal wavelet basis and its associated analysis and synthesis matrices.

1. *Select the scaling functions* $\mathbf{\Phi}^j(x)$, *either explicitly, or via a subdivision scheme.* This choice determines the nested approximation spaces V^j and the synthesis filters P^j.

2. *Select an inner product defined on the functions in* $V^0, V^1, \ldots$. This choice determines the L^2 norm and defines orthogonality. Although the standard inner product is the com-

mon choice, a weighted inner product can be used to capture a measure of error that is meaningful in the context of the application at hand.

3. *Select wavelet synthesis matrices Q^j that satisfy $(P^j)^{\mathrm{T}}[\langle \Phi^j \mid \Phi^j \rangle]Q^j = 0$.* This choice determines the wavelets $\Psi^j(x)$ that span the spaces W^j. Together, the synthesis filters P^j and Q^j determine the analysis filters A^j and B^j by equation (7.10).

7.4 Biorthogonal wavelets

So far, we have defined wavelets as bases for orthogonal complements. Requiring orthogonality between the spaces V^j and W^j is convenient, especially when a fully orthonormal basis is used. However, it does have some undesirable consequences:

- Semiorthogonal constructions typically guarantee that P^j and Q^j are sparse and therefore that reconstruction can be done in linear time. However, there is no guarantee that A^j and B^j will be sparse, so analysis may take quadratic time. In Section 7.3.2 we described how LU decomposition could be used to maintain linear-time complexity for spline-wavelet analysis, but this trick unfortunately doesn't work when applying analysis to higher-dimensional functions such as surfaces.

- Adding a single orthogonal or semiorthogonal wavelet to a function made up of coarser scaling functions generally introduces many "new vertices." In some applications, it is desirable for each wavelet to introduce just a single new vertex while merely perturbing the existing vertices.

- There is no known general technique for improving an existing orthogonal or semiorthogonal wavelet basis while still satisfying the constraints of the original construction.

Fortunately, it is possible to define wavelets that, although not orthogonal to scaling functions, still have many of the desirable properties of the semiorthogonal wavelets described in the previous section. Multiresolution bases of this type are called *biorthogonal wavelets*, and they were first developed by Cohen et al. [20].

Before motivating the term *biorthogonality*, we first note that much of the material developed earlier in this chapter is sufficiently general to apply to the biorthogonal setting. The principle difference is that we no longer require Q^j to be in the null space of $(P^j)^{\mathrm{T}} [\langle \Phi^j \mid \Phi^j \rangle]$. Instead, all that is necessary is that the matrix $[\, P^j \mid Q^j \,]$ be invertible, implying only that the analysis matrices A^j and B^j exist.

The additional trick in building a biorthogonal basis is in arranging for the analysis and synthesis matrices to be sparse in order to allow for fast filter-bank decomposition and reconstruction. As we'll see, the *lifting scheme*, described in Section 7.4.4, can be used to build biorthogonal bases with a number of nice properties while maintaining this sparsity.

7.4.1 Duals and biorthogonality

Before we define biorthogonality, let's consider the more familiar situation of orthogonality. Specifically, consider an orthonormal basis $u(x) = [u_1(x) \quad u_2(x) \quad \cdots]$ and the problem of finding coefficients c_i that represent a given function $f(x)$ in the span of $u(x)$:

$$f(x) = \sum_i c_i u_i(x)$$

An arbitrary coefficient c_j can be determined directly by computing the inner product of both sides of this equation with $u_j(x)$:

$$\langle f \mid u_j \rangle = \sum_i c_i \langle u_i \mid u_j \rangle$$

$$= \sum_i c_i \delta_{i,j}$$

$$= c_j$$

Suppose now that $u(x)$ is *not* an orthonormal basis. Although $\langle f \mid u_j \rangle \neq c_j$, we can still hope to find a new collection of functions $\tilde{u}(x) = [\tilde{u}_1(x) \quad \tilde{u}_2(x) \quad \cdots]$ such that

$$c_j = \langle f \mid \tilde{u}_j \rangle \tag{7.15}$$

One of the fundamental results of linear algebra states that for any finite-dimensional basis, such a collection of functions always exists. In this context, the original functions $u(x)$ are called the *primal basis*, and the new functions $\tilde{u}(x)$ are called the *dual basis*. If equation (7.15) is to hold for any function $f(x)$ in the span of $u(x)$, then it follows that $\langle u_i \mid \tilde{u}_j \rangle = \delta_{i,j}$ (it's a nice little exercise to prove this). Equivalently, we can define the dual basis corresponding to a primal basis $u(x)$ to be a set of functions $\tilde{u}(x)$ satisfying the following:

$$[\langle u \mid \tilde{u} \rangle] = I$$

Note that if $\boldsymbol{u}(x)$ is orthonormal, then it is *self-dual*; that is, $\tilde{\boldsymbol{u}}(x) = \boldsymbol{u}(x)$.

Dual basis functions are central to the definition of a biorthogonal wavelet construction. First, let $\tilde{\boldsymbol{\Phi}}^j(x)$ and $\tilde{\boldsymbol{\Psi}}^j(x)$ be the duals corresponding to $\boldsymbol{\Phi}^j(x)$ and $\boldsymbol{\Psi}^j(x)$, respectively. Thus,

$$[\langle \boldsymbol{\Phi}^j \mid \tilde{\boldsymbol{\Phi}}^j \rangle] = \boldsymbol{I} \quad \text{and} \quad [\langle \boldsymbol{\Psi}^j \mid \tilde{\boldsymbol{\Psi}}^j \rangle] = \boldsymbol{I} \tag{7.16}$$

A biorthogonal wavelet basis is one in which the primal scaling functions are orthogonal to the dual wavelets and the primal wavelets are orthogonal to the dual scaling functions. In other words, *biorthogonal wavelets* must satisfy the conditions

$$\left. \begin{array}{l} \langle \phi_k^j \mid \tilde{\psi}_\ell^j \rangle = 0 \\ \langle \psi_k^j \mid \tilde{\phi}_\ell^j \rangle = 0 \end{array} \right\} \quad \text{for all } j, k, \text{ and } \ell$$

We can write this definition equivalently as

$$[\langle \boldsymbol{\Phi}^j \mid \tilde{\boldsymbol{\Psi}}^j \rangle] = \boldsymbol{0} \quad \text{and} \quad [\langle \boldsymbol{\Psi}^j \mid \tilde{\boldsymbol{\Phi}}^j \rangle] = \boldsymbol{0} \tag{7.17}$$

Combining equations (7.16) and (7.17) yields a concise statement of both duality and biorthogonality:

$$\left[\left\langle \, [\boldsymbol{\Phi}^j \mid \boldsymbol{\Psi}^j] \; \middle| \; [\tilde{\boldsymbol{\Phi}}^j \mid \tilde{\boldsymbol{\Psi}}^j] \, \right\rangle \right] = \left[\begin{array}{c|c} [\langle \boldsymbol{\Phi}^j \mid \tilde{\boldsymbol{\Phi}}^j \rangle] & [\langle \boldsymbol{\Phi}^j \mid \tilde{\boldsymbol{\Psi}}^j \rangle] \\ \hline [\langle \boldsymbol{\Psi}^j \mid \tilde{\boldsymbol{\Phi}}^j \rangle] & [\langle \boldsymbol{\Psi}^j \mid \tilde{\boldsymbol{\Psi}}^j \rangle] \end{array} \right] = \boldsymbol{I} \tag{7.18}$$

7.4.2 Defining the duals through subdivision

It turns out that dual functions are lurking whenever the synthesis process can be inverted. The primal functions are of course just the scaling functions $\boldsymbol{\Phi}^j(x)$ and wavelets $\boldsymbol{\Psi}^j(x)$ defined by the synthesis matrices. Since these functions can be defined through subdivision, it is natural to wonder whether their duals, $\tilde{\boldsymbol{\Phi}}^j(x)$ and $\tilde{\boldsymbol{\Psi}}^j(x)$, can also be defined through subdivision. Indeed, this turns out to be the case, at least up to a constant scale factor, as we now show.

Suppose we have the synthesis matrices $\boldsymbol{P}^j$ and $\boldsymbol{Q}^j$ defining a set of primal scaling functions and wavelets according to

$$[\ \Phi^{j-1}\ |\ \Psi^{j-1}\ ] = \Phi^j\ [\ P^j\ |\ Q^j\ ] \tag{7.19}$$

We would like to find matrices $\tilde{P}^j$ and $\tilde{Q}^j$ that define dual scaling functions and wavelets through subdivision:

$$[\ \tilde{\Phi}^{j-1}\ |\ \tilde{\Psi}^{j-1}\ ] = \tilde{\Phi}^j\ [\ \tilde{P}^j\ |\ \tilde{Q}^j\ ] \tag{7.20}$$

Before continuing, we need to address a somewhat subtle issue relating to normalization. The simplest subdivision schemes are stationary, meaning that the scaling functions are almost everywhere dilates and translates of a single function. Stationary subdivision, by definition, imposes a normalization in which the height of a scaling function ϕ_i^j remains constant over all j. This normalization is at odds with the normalization required by biorthogonality: as the supports of the primals and duals become narrower, the heights of the duals would have to grow for their inner product with the primals to remain 1. It is therefore easier to use subdivision to characterize functions that are biorthogonal up to an appropriate constant, with the constant related to the rate at which supports shrink during subdivision. We'll denote the unnormalized duals by $\overline{\phi}$ and $\overline{\psi}$, and the subdivision matrices that construct them $\overline{P}$ and $\overline{Q}$.

In one dimension (that is, for functions of one parameter, such as curves), supports shrink by a factor of 2 each time the resolution index j increases by 1. In two dimensions (that is, for surfaces), supports shrink by a factor of 2 in each direction, for an overall shrinkage factor of 4. In n dimensions, the shrinkage factor is 2^n. We will therefore construct unnormalized duals such that the inner product of a scaling function or wavelet at resolution j with its dual is 2^{-jn}. The biorthogonality condition for the unnormalized duals at resolution j can be written compactly as

$$\left[\left\langle\ [\ \Phi^j\ |\ \Psi^j\ ]\ \middle|\ [\ \overline{\Phi}^j\ |\ \overline{\Psi}^j\ ]\ \right\rangle \right] = 2^{-jn}\,I \tag{7.21}$$

Once the unnormalized duals at resolution j are found, they can be properly normalized by multiplying each of them by 2^{jn}. For example, scaling function and wavelet coefficients for a function $f(x)$ can be computed using the unnormalized dual scaling function $\overline{\phi}_i^j(x)$ and wavelet $\overline{\psi}_i^j(x)$ as follows:

$$c_i^j = 2^{jn}\langle f\ |\ \overline{\phi}_i^j \rangle$$
$$d_i^j = 2^{jn}\langle f\ |\ \overline{\psi}_i^j \rangle$$

The first step in determining the dual subdivision matrices $\overline{P}^j$ and $\overline{Q}^j$ is to rewrite equation (7.21) using superscript $j-1$ instead of j:

$$\left[\left\langle \; [\Phi^{j-1} \mid \Psi^{j-1}] \; \middle| \; [\overline{\Phi}^{j-1} \mid \overline{\Psi}^{j-1}] \; \right\rangle \right] = 2^{-(j-1)n} I$$

Next, we substitute equations (7.19) and (7.20) into the previous equation, and use the properties of inner product matrices from equation (7.12) to get

$$[P^j \mid Q^j]^{\mathrm{T}} [\langle \Phi^j \mid \overline{\Phi}^j \rangle] [\overline{P}^j \mid \overline{Q}^j] = 2^{-(j-1)n} I$$

We can use equation (7.21) again to show that the middle term on the left-hand side is just $2^{-jn}I$, so we have

$$[\overline{P}^j \mid \overline{Q}^j]^{\mathrm{T}} [P^j \mid Q^j] = 2^n I \tag{7.22}$$

In other words, up to a factor of 2^n, the primal and dual synthesis matrices are inverse transposes of one another. Recall the similar equation relating the analysis and synthesis matrices:

$$\left[\frac{A^j}{B^j} \right] [P^j \mid Q^j] = I$$

Comparing these equations, it is evident that

$$\overline{P}^j = 2^n (A^j)^{\mathrm{T}} \tag{7.23}$$
$$\overline{Q}^j = 2^n (B^j)^{\mathrm{T}} \tag{7.24}$$

Now we have a new interpretation for the analysis matrices A^j and B^j: with the appropriate constant factor in front, they define a new subdivision scheme with its own scaling functions and wavelets. The scaling functions and wavelets of that new scheme are dual to the scaling functions and wavelets of the original scheme defined by P^j and Q^j.

Remark: Strictly speaking, we have shown that if duals generated through subdivision exist, then their synthesis matrices must be given by equations (7.23) and (7.24). We would really like to know that if we use the matrices in equations (7.23) and (7.24) as subdivision matrices, then the resulting functions are truly dual to the original scaling functions and wavelets. This relation seems to hold true in practice, but appears to be quite difficult to prove. ∎

The dual functions $\bar{\phi}_i^j(x)$ and $\bar{\psi}_i^j(x)$ don't crop up very often—their main use is in finding the coefficients to use with the primal basis functions in order to represent some given function. We can nonetheless use the dual refinement matrices $\bar{P}^j$ and $\bar{Q}^j$ to compute what the dual basis functions look like. To construct a dual function $\bar{\phi}_i^j$ or $\bar{\psi}_i^j$, we start with a single 1 embedded in a sequence of zeros and multiply by $\bar{P}^j$ or $\bar{Q}^j$, respectively. Then we continue to subdivide by applying $\bar{P}^{j+1}$, $\bar{P}^{j+2}$, and so on. Remember that the resulting function should be multiplied by 2^{jn} to obtain a normalized dual $\tilde{\phi}_i^j$ or $\tilde{\psi}_i^j$ (whose inner product with the corresponding primal function is 1).

7.4.3 Single-knot wavelets

To demonstrate that it is actually possible to develop a biorthogonal construction that has linear-time analysis and synthesis, we will now look in detail at the case of efficiently decomposing piecewise-linear functions. In order to avoid the complications that arise at the ends of a bounded interval, we'll focus our attention on the interior of an interval. Thus, the wavelets we develop here are equally well suited to a periodic domain or to the unbounded real line. The wavelet basis that we develop in this section will be particularly useful for hierarchically decomposing closed polygonal contours when we discuss multiresolution tiling in Chapter 9.

The first step in the construction is to take as the scaling functions for V^j the *hat functions* centered over the integers and to take as the scaling functions for V^{j-1} hat functions that are twice as wide, centered above the even integers, as shown in Figure 7.5(a) and (b). A very simple way to obtain a linear-time biorthogonal construction is to simply use the hat functions in V^j centered on the odd integers as the wavelets for W^{j-1}, as shown in Figure 7.5(c). These wavelets have been dubbed *lazy wavelets* by Wim Sweldens [122], because nothing needs to be done to compute them—they're just a subset of the scaling functions.

To see just how simple the analysis and synthesis processes are for the lazy wavelet basis, we'll examine the matrices that are involved. The scaling function refinement matrix P_{lazy}^j has columns that consist of only three nonzero terms, $(\frac{1}{2}, 1, \frac{1}{2})$, as a wide hat function can be constructed from three narrow hat functions. The columns of the lazy wavelet refinement matrix Q_{lazy}^j contain all zeros except for a single 1, because each lazy wavelet is a narrow hat function. An example will help clarify the sparse structure of these matrices, so here are the lazy wavelet synthesis matrices corresponding to $j = 3$, with dots taking the place of zeros:

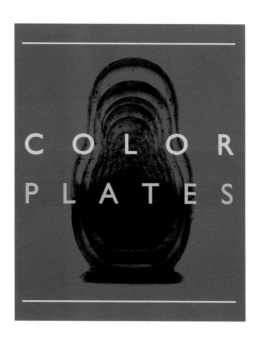

COLOR
PLATES

(a)

(b)

(c)

(d)

PLATE 1 Compressing a color image (from Renoir's painting "La Promenade") using the nonstandard Haar basis and the L^2 norm: (a) original image; (b) 5% L^2 error (6:1 compression); (c) 10% L^2 error (42:1 compression); (d) 15% L^2 error (222:1 compression). (See Chapter 3: Image Compression.)

(a) (b)

(c) (d)

PLATE 2 Adding makeup and a glint in the eye to da Vinci's "Mona Lisa": (a) Mona Lisa; (b) adding makeup at higher resolution; (c) zooming in a little closer for the glint; (d) Mona ready to step out. (See Chapter 4: Multiresolution Image Editing.)

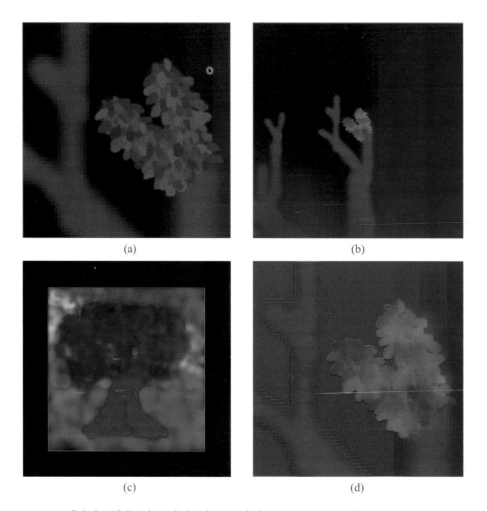

PLATE 3 Painting "in" and "under" at low resolutions: (a) close-up of leaves on a tree; (b) painting "in" with fall colors; (c) painting "under" with sky and grass; (d) modified close-up view. (See Chapter 4: Multiresolution Image Editing.)

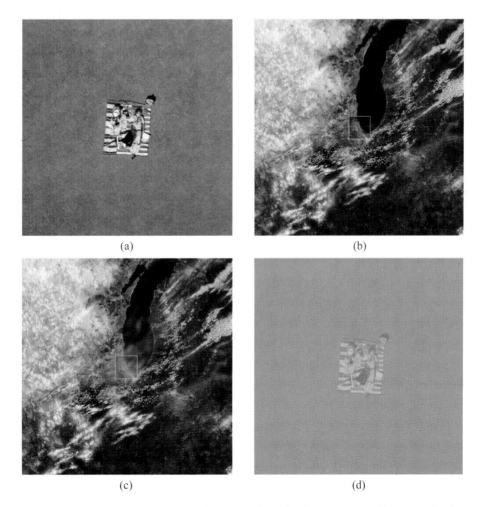

(a)

(b)

(c)

(d)

PLATE 4 Adding smog to an image with a range in scale of 100,000 to 1: (a) a sunny day in the park; (b) zooming out by a factor of 100,000; (c) adding smog over Chicago; (d) smog affects the park view. (See Chapter 4: Multiresolution Image Editing.)

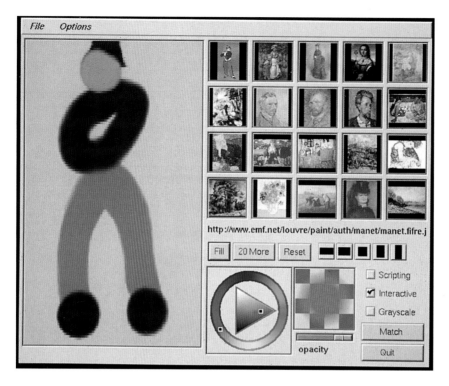

PLATE 5 The image querying application. The user paints a query in the large rectangular window, and the 20 highest-ranked targets appear in the small windows on the right. To avoid copyright infringements, the database for this example contains only 96 images (all created by artists who have been dead more than 75 years). Because the database is so limited, only the intended target (in the upper-left small window) appears to match the query very closely. (See Chapter 5: Multiresolution Image Querying.)

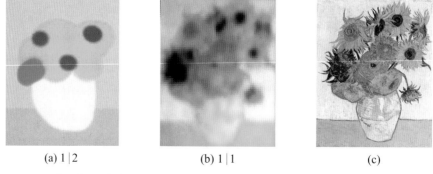

(a) 1 | 2   (b) 1 | 1   (c)

PLATE 6 Queries and their target: (a) a query painted from memory; (b) a scanned query; (c) their intended target. Below the queries, the L^q ranks of the intended target are shown for two databases of sizes 1093 | 20,558. (See Chapter 5: Multiresolution Image Querying.)

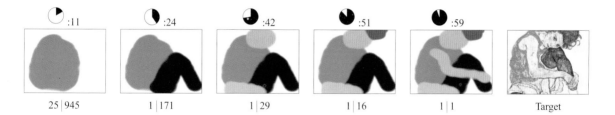

| 25 \| 945 | 1 \| 171 | 1 \| 29 | 1 \| 16 | 1 \| 1 | Target |

PLATE 7 Progression of an interactive query. Above each partially formed query is the actual time (in seconds) at which the snapshot was taken. Below each query are the L^q ranks of the intended target for databases of sizes 1093 \| 20,558. (See Chapter 5: Multiresolution Image Querying.)

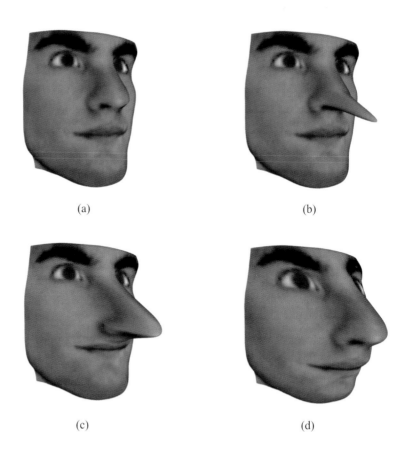

(a)

(b)

(c)

(d)

PLATE 8 A tensor-product spline surface, manipulated at different levels of detail. The original surface (a) is changed at a narrow scale (b), an intermediate scale (c), and a broad scale (d). (See Chapter 10: Surface Wavelets.)

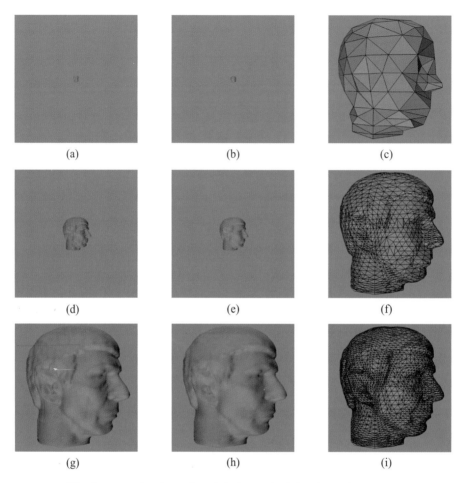

PLATE 9 Wavelet approximations of a polyhedron. The left column shows Gouraud-shaded views of the full-resolution model with 16,386 vertices and 32,768 triangles. Compare the distant view in (a) with a compressed wavelet approximation in (b) consisting of only 105 wavelets and 240 triangles, shown in a flat-shaded close-up in (c). Similarly, the mid-range view in (d) is approximated in (e) by a compressed model with 1,966 wavelets and 4,272 triangles, shown in (f). Finally, the close-up view in (g) is approximated in (h) by a model with 5,054 wavelets and 10,704 triangles, shown in (i). (See Chapter 11: Surface Applications.)

PLATE 10 Compressed wavelet approximations of a bunny model: (a) 952 wavelet coefficients; (b) 2,268 wavelet coefficients; (c) 18,636 wavelet coefficients. (See Chapter 11: Surface Applications.)

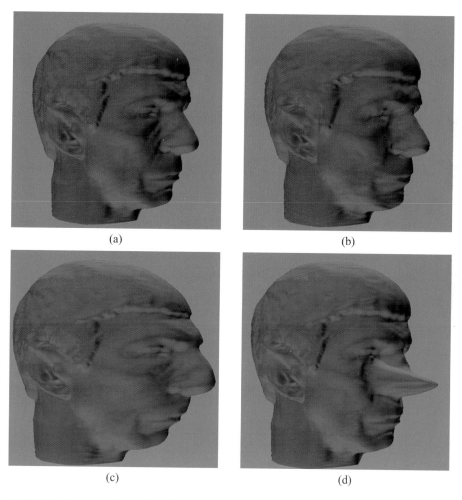

PLATE 11 Compression and multiresolution editing of a smooth surface. The original smooth surface (a) can be compressed to 16% of its wavelet coefficients (b). Editing the original surface can take place at a coarse level (c) or a finer level (d). (See Chapter 11: Surface Applications.)

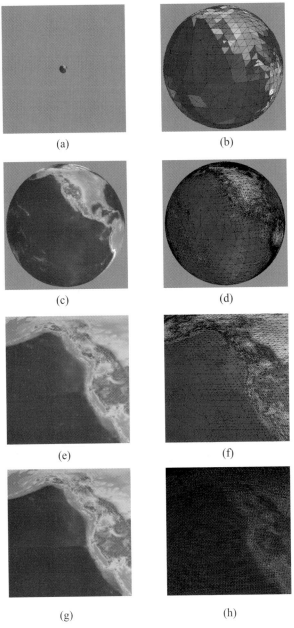

PLATE 12 Approximating color as a function over the sphere. The last row shows the full-resolution model. (See Chapter 11: Surface Applications.)

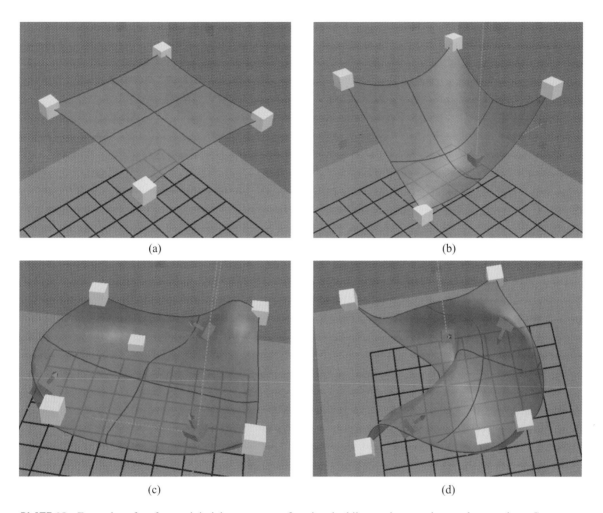

PLATE 13 Examples of surfaces minimizing an energy functional while meeting user-imposed constraints. Cyan cubes represent positional constraints, while purple cubes represent constraints on tangents and normals. (See Chapter 12: Variational Modeling.)

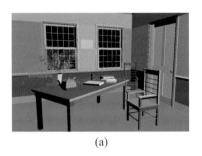 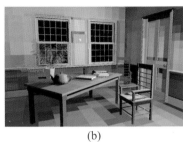 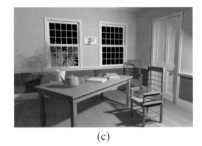

| (a) | (b) | (c) |

PLATE 14 Radiosity solutions computed by a wavelet radiosity algorithm: (a) initial solution in space V^0; (b) refined solution in space V^7; (c) refined solution with final gathering step. (See Chapter 13: Global Illumination.)

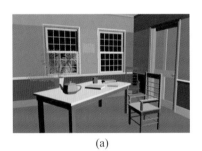 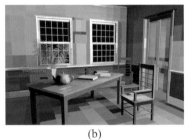 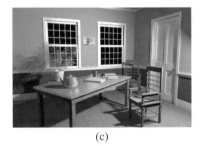

| (a) | (b) | (c) |

PLATE 15 Radiance solutions computed by a wavelet radiance algorithm: (a) initial solution in space V^0; (b) refined solution in space V^6; (c) refined solution with final gathering step. (See Chapter 13: Global Illumination.)

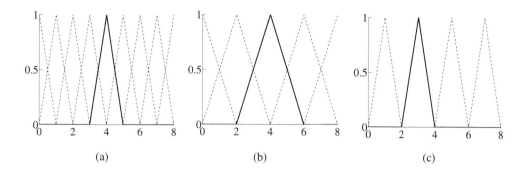

FIGURE 7.5 (a) Piecewise-linear scaling functions in V^j; (b) scaling functions in V^{j-1}; (c) lazy wavelets in W^{j-1}.

$$
[\, P_{\text{lazy}}^3 \mid Q_{\text{lazy}}^3 \,] = \frac{1}{2}
\left[
\begin{array}{ccccc|ccccc}
2 & \cdot & \cdot & \cdot & \cdot & \cdot & \cdot & \cdot & \cdot \\
1 & 1 & \cdot & \cdot & \cdot & 2 & \cdot & \cdot & \cdot \\
\cdot & 2 & \cdot & \cdot & \cdot & \cdot & \cdot & \cdot & \cdot \\
\cdot & 1 & 1 & \cdot & \cdot & \cdot & 2 & \cdot & \cdot \\
\cdot & \cdot & 2 & \cdot & \cdot & \cdot & \cdot & \cdot & \cdot \\
\cdot & \cdot & 1 & 1 & \cdot & \cdot & \cdot & 2 & \cdot \\
\cdot & \cdot & \cdot & 2 & \cdot & \cdot & \cdot & \cdot & \cdot \\
\cdot & \cdot & \cdot & 1 & 1 & \cdot & \cdot & \cdot & 2 \\
\cdot & \cdot & \cdot & \cdot & 2 & \cdot & \cdot & \cdot & \cdot \\
\end{array}
\right]
$$

The corresponding analysis matrices, which are given by the inverse of $[\, P_{\text{lazy}}^3 \mid Q_{\text{lazy}}^3 \,]$, are also sparse:

$$
\left[
\begin{array}{c}
A_{\text{lazy}}^3 \\
\hline
B_{\text{lazy}}^3
\end{array}
\right]
= \frac{1}{2}
\left[
\begin{array}{ccccccccc}
2 & \cdot & \cdot & \cdot & \cdot & \cdot & \cdot & \cdot & \cdot \\
\cdot & \cdot & 2 & \cdot & \cdot & \cdot & \cdot & \cdot & \cdot \\
\cdot & \cdot & \cdot & \cdot & 2 & \cdot & \cdot & \cdot & \cdot \\
\cdot & \cdot & \cdot & \cdot & \cdot & \cdot & 2 & \cdot & \cdot \\
\cdot & \cdot & \cdot & \cdot & \cdot & \cdot & \cdot & \cdot & 2 \\
\hline
-1 & 2 & -1 & \cdot & \cdot & \cdot & \cdot & \cdot & \cdot \\
\cdot & \cdot & -1 & 2 & -1 & \cdot & \cdot & \cdot & \cdot \\
\cdot & \cdot & \cdot & \cdot & -1 & 2 & -1 & \cdot & \cdot \\
\cdot & \cdot & \cdot & \cdot & \cdot & \cdot & -1 & 2 & -1 \\
\end{array}
\right]
$$

We've now established that sparse matrices P_{lazy}^j, Q_{lazy}^j, A_{lazy}^j, and B_{lazy}^j exist such that

$$[\, P^j_{\text{lazy}} \mid Q^j_{\text{lazy}} \,] \begin{bmatrix} A^j_{\text{lazy}} \\ \hline B^j_{\text{lazy}} \end{bmatrix} = I$$

Although lazy wavelets make an interesting example, they are not of much practical use. To see why, consider what happens as we perform analysis on a given piecewise-linear function. A coarse approximation is formed from a finer one by simply leaving out the odd-numbered vertices. Therefore, the coarsest approximation depends on only a few of the original vertices, and it may be a very poor approximation of the overall shape of the original function.

The lazy wavelets can be improved by making them "more orthogonal" to V^{j-1}. We do this by subtracting off some amount of nearby coarse scaling functions from each wavelet. The i-th such wavelet would therefore have the form

$$\psi^{j-1}_i(x) = \phi^j_{2i+1}(x) - \sum_k s^j_{k,i}\, \phi^{j-1}_{i+k}(x)$$

where the summation over k is limited to only a few terms. The coefficients $s^j_{k,i}$ can be determined by requiring that $\psi^{j-1}_i(x)$ be as close as possible to orthogonal to the scaling functions in V^{j-1}. Satisfying this condition requires solving, in a least-squares sense [47], the overdetermined system of equations

$$\langle\, \psi^{j-1}_i \mid \phi^{j-1}_{i'} \,\rangle = 0 \tag{7.25}$$

for all i' such that the support of $\phi^{j-1}_{i'}(x)$ overlaps with $\psi^{j-1}_i(x)$. Piecewise-linear wavelets satisfying equation (7.25) have been dubbed *single-knot wavelets* by Joe Warren [126] because when a wavelet is added to a function with vertices at the even integers, only a single new vertex (or knot) needs to be placed at an odd integer. One such wavelet is shown in Figure 7.6(a) for the case where the summation over k is restricted to four terms corresponding to $k = -1, 0, 1, 2$. The values for s as well as the analysis and synthesis sequences for this scheme are listed below:

$$s = \tfrac{1}{258}(-19, 81, 81, -19)$$

$$p = \tfrac{1}{2}(1, 2, 1)$$

$$q = \tfrac{1}{516}(19, 38, -62, -162, 354, -162, -62, 38, 19)$$

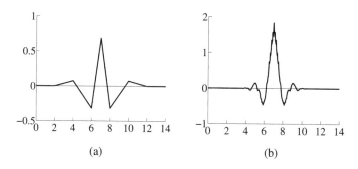

(a)   (b)

FIGURE 7.6   (a) One of the single-knot wavelets; (b) a dual scaling function corresponding to the single-knot construction.

$$a = \tfrac{1}{516}\,(19,\,-38,\,-62,\,162,\,354,\,162,\,-62,\,-38,\,19)$$
$$b = \tfrac{1}{2}\,(-1,\,2,\,-1)$$

The dual functions that are defined through subdivision using the a and b sequences can sometimes look pretty bizarre, even when the primal functions are simple. As an example, Figure 7.6(b) shows a dual scaling function corresponding to the single-knot wavelet construction. The refinement sequence for the dual scaling functions is twice the analysis sequence a given above. This plot (like any plot we've generated through subdivision) was created using the iterative splitting and averaging method described in Section 6.4.

The construction of the single-knot wavelet can be summarized by grouping the coefficients $s_{k,i}^{j}$ into a matrix S^{j}. The single-knot (sk) synthesis matrix is then constructed from the lazy wavelet synthesis matrix $[\,P_{\text{lazy}}^{j}\mid Q_{\text{lazy}}^{j}\,]$ by defining

$$[\,P_{\text{sk}}^{j}\mid Q_{\text{sk}}^{j}\,] := [\,P_{\text{lazy}}^{j}\mid Q_{\text{lazy}}^{j} - P_{\text{lazy}}^{j}\,S^{j}\,]$$

Single-knot synthesis is fast since all of the terms on the right-hand side of the relation above are sparse (including S^{j}). Single-knot analysis is also fast, because the analysis matrices are also sparse:

$$\begin{bmatrix} A_{\text{sk}}^{j} \\ \hline B_{\text{sk}}^{j} \end{bmatrix} = [\,P_{\text{lazy}}^{j}\mid Q_{\text{lazy}}^{j} - P_{\text{lazy}}^{j}\,S^{j}\,]^{-1} = \begin{bmatrix} A_{\text{lazy}}^{j} + S^{j}\,B_{\text{lazy}}^{j} \\ \hline B_{\text{lazy}}^{j} \end{bmatrix}$$

7.4.4 Designing biorthogonal wavelets: lifting

The single-knot wavelet construction given above, first reported by Lounsbery et al. [75], was recognized by Sweldens [122] as a special case of a general transformation, which he dubbed *lifting*. Lifting is an operation that turns a biorthogonal scheme defined by matrices P^j, Q^j, A^j, and B^j into a new biorthogonal scheme defined by

$$[\, P_{\text{lift}}^j \mid Q_{\text{lift}}^j \,] = [\, P^j \mid Q^j - P^j S^j \,] \quad \text{and} \quad \begin{bmatrix} A_{\text{lift}}^j \\ B_{\text{lift}}^j \end{bmatrix} = \begin{bmatrix} A^j + S^j B^j \\ B^j \end{bmatrix} \tag{7.26}$$

The lifted scheme is biorthogonal for *any* choice of the matrix S^j, not just for those matrices resulting from least-squares projections. By choosing S^j in different ways, it is possible to build bases with various desirable attributes, such as increased orthogonality, higher vanishing moments, and so on (see the papers by Sweldens [122] and Schröder and Sweldens [109] for details).

The lifting operation described by equation (7.26) leaves the matrices P^j and B^j unchanged, but it modifies the other two matrices Q^j and A^j. Thus, lifting changes the primal wavelets (through the change to Q^j), the dual scaling functions (through the change to A^j), *and* the dual wavelets since they are linear combinations of the modified dual scaling functions.

It is also possible to perform *dual lifting* [122], in which an initial scheme defined by matrices P^j, Q^j, A^j, and B^j is replaced by a scheme defined by the matrices

$$[\, P_{\widetilde{\text{lift}}}^j \mid Q_{\widetilde{\text{lift}}}^j \,] = [\, P^j + S^j Q^j \mid Q^j \,] \quad \text{and} \quad \begin{bmatrix} A_{\widetilde{\text{lift}}}^j \\ B_{\widetilde{\text{lift}}}^j \end{bmatrix} = \begin{bmatrix} A^j \\ B^j - A^j S^j \end{bmatrix}$$

Dual lifting leaves the dual scaling functions unchanged, but modifies the dual wavelets, as well as the primal scaling functions and wavelets. Schröder and Sweldens [109] give examples of wavelet bases constructed using dual lifting.

In summary, the steps involved in constructing a biorthogonal wavelet basis and its associated analysis and synthesis matrices are straightforward:

1. *Select a simple biorthogonal construction with known sparse synthesis and analysis matrices P^j, Q^j, A^j, and B^j.*

2. *Apply lifting or dual lifting to modify the synthesis and analysis matrices in a way that improves the biorthogonal construction.*

TABLE 7.1   Summary of refinement, analysis, synthesis, and invertibility equations.

Process	Definition	Block-Matrix Form	Equation
Refinement	$\Phi^{j-1}(x) = \Phi^j(x)\,P^j$ $\Psi^{j-1}(x) = \Phi^j(x)\,Q^j$	$[\ \Phi^{j-1}\ \|\ \Psi^{j-1}\ ] = \Phi^j\,[\ P^j\ \|\ Q^j\ ]$	(7.3)
Analysis	$c^{j-1} = A^j\,c^j$ $d^{j-1} = B^j\,c^j$	$\begin{bmatrix} c^{j-1} \\ d^{j-1} \end{bmatrix} = \begin{bmatrix} A^j \\ B^j \end{bmatrix} c^j$	(7.4) (7.5)
Synthesis	$c^j = P^j\,c^{j-1} + Q^j\,d^{j-1}$	$c^j = [\ P^j\ \|\ Q^j\ ] \begin{bmatrix} c^{j-1} \\ d^{j-1} \end{bmatrix}$	(7.6)
Invertibility	$A^j\,P^j = B^j\,Q^j = I$ $A^j\,Q^j = B^j\,P^j = 0$ $P^j\,A^j + Q^j\,B^j = I$	$\begin{bmatrix} A^j \\ B^j \end{bmatrix} = [\ P^j\ \|\ Q^j\ ]^{-1}$	(7.10)

7.5 Summary

The framework for multiresolution analysis laid out at the beginning of this chapter is very general. It subsumes a variety of wavelet constructions, including orthogonal, semiorthogonal, and biorthogonal wavelet bases constructed on any compact domain. The most important equations of matrix-based multiresolution analysis are summarized in Table 7.1.

The conditions that define orthogonal, semiorthogonal, and biorthogonal constructions of wavelet bases are summarized in Table 7.2. The third column of the table helps illustrate the fact that orthogonality is a special case of semiorthogonality, which in turn is a special case of biorthogonality. Deciding whether to use an orthogonal, semiorthogonal, or biorthogonal wavelet basis is a difficult task—each construction has its own advantages and disadvantages. Orthogonal bases have the useful property that L^2 error metrics are easily computed from basis function coefficients; however, very few function spaces admit constructions of orthogonal wavelets. A semiorthogonal construction offers the opportunity to use compactly supported, symmetric wavelets while maintaining orthogonality between wavelets and scaling functions. Biorthogonal wavelets allow the most flexibility, as lifting can be used to improve such properties as locality of support and degree of orthogonality.

TABLE 7.2   Summary of conditions defining orthogonal, semiorthogonal, and biorthogonal wavelets.

Property	Basis Function Constraints	Matrix Constraints	Equation
Orthogonality	$[\langle \Phi^j \mid \Phi^j \rangle] = I$ $[\langle \Psi^j \mid \Psi^j \rangle] = I$ $[\langle \Phi^j \mid \Psi^j \rangle] = 0$	$[\,P^j \mid Q^j\,]$ invertible and orthogonal	(7.11)
Semiorthogonality	$[\langle \Phi^j \mid \Psi^j \rangle] = 0$	$[\,P^j \mid Q^j\,]$ invertible and $(P^j)^{\mathrm{T}}[\langle \Phi^j \mid \Phi^j \rangle]\,Q^j = 0$	(7.13)
Biorthogonality	$[\langle \Phi^j \mid \tilde{\Phi}^j \rangle] = I$ $[\langle \Psi^j \mid \tilde{\Psi}^j \rangle] = I$ $[\langle \Phi^j \mid \tilde{\Psi}^j \rangle] = 0$ $[\langle \Psi^j \mid \tilde{\Phi}^j \rangle] = 0$	$[\,P^j \mid Q^j\,]$ invertible	(7.16) (7.17)

MULTIRESOLUTION CURVES

1. Related curve representations — 2. Smoothing a curve — 3. Editing a curve —
4. Scan conversion and curve compression

Curves play a fundamental role in many graphics applications, including computer-aided design, in which cross-sectional curves are frequently used in the specification of surfaces; keyframe animation, in which curves are used to control parameter interpolation; three-dimensional modeling and animation, in which "backbone" curves are manipulated to specify object deformations; graphic design, in which curves are used to describe regions of constant color or texture; font design, in which curves represent the outlines of characters; and pen-and-ink illustration, in which curves are the basic elements of the finished piece. All of these applications benefit from a representation for curves that allows flexible editing, smoothing, and scan conversion. In particular, a good representation for curves should support

- continuous levels of smoothing, in which undesirable features are removed from a curve (see Figure 8.1, page 112)

- the ability to change the overall "sweep" of a curve while maintaining its fine details, or "character" (see Figure 8.2, page 113)

- the ability to edit a curve at any continuous level of detail, allowing an arbitrary portion of the curve to be affected through direct manipulation (see Figure 8.3, page 114)

- the ability to change a curve's character without affecting its overall sweep (see Figure 8.5, page 119)

- curve approximation, or "fitting," within a guaranteed error tolerance, for scan conversion and other applications (see Figure 8.7, page 123)

In this chapter, we describe how a *multiresolution curve* representation can provide a single, unified framework for addressing all of these issues. Multiresolution curves, which were first presented by Finkelstein and Salesin [37], are built from the B-spline wavelets developed in Section 7.3.2. The multiresolution representation of a curve requires no extra storage beyond that of the original m B-spline control points, and the algorithms that use it are both simple and fast, typically linear in m.

8.1   Related curve representations

The usefulness of a multiresolution curve representation lies in the convenience with which a wide variety of operations can be performed. A number of other representations for curves have been developed to facilitate one or more of these operations. In this section we survey some of these techniques.

Forsey and Bartels [39] employ hierarchical B-splines to address the problem of editing the overall form of a surface while maintaining its details. Their original formulation requires the user to design an explicit hierarchy into the model. In a later work [41], they describe a method for recursively fitting a hierarchical surface to a set of data by first finding a coarse approximation and then refining in areas where the approximation needs more detail. This construction is similar in spirit to the filter-bank process used in multiresolution analysis, as described in Section 7.1.2. However, when using hierarchical B-splines an infinite number of representations are possible for a given curve or surface, whereas the multiresolution curve representation of the same shape is unique. Fowler [42] and Welch and Witkin [130] also describe methods in which editing can be performed over narrower or broader regions of a surface. Neither of these works, however, attempts to preserve high-resolution details when low-resolution editing occurs.

A variety of techniques exist for smoothing curves and surfaces. Among these are algorithms that minimize an energy norm; these are surveyed in Hoschek and Lasser [61]. One example is the work of Celniker and Gossard [9], in which a fairness functional is applied to hand-drawn curves as well as to surfaces. These minimization techniques fall into the category of variational modeling, which we will discuss further in the context of wavelets in Chapter 12. The method of smoothing we describe in this chapter simply minimizes a least-squares error measure—a much simpler approach that allows the amount of smoothing to be varied continuously.

Many schemes for approximating a curve within a specified error tolerance have also been explored [1, 76, 95, 106]. Most of this research has centered on various forms of knot removal for representing curves efficiently with nonuniform B-splines. In Section 8.4, we look at a slightly different approach to curve approximation. We describe how to approximate a curve well using a small number of Bézier segments—a very practical application, since these segments are the standard representation for curves in PostScript, the most prevalent page description language.

In the remainder of this chapter, we discuss how the particular form of multiresolution analysis developed for B-splines in Section 7.3.2 can be applied directly to curves. The use of spline wavelets leads to efficient algorithms for smoothing, editing, and scan-converting curves.

8.2 Smoothing a curve

In this section, we address the following problem: Given a curve $\gamma(t)$ with m control points $c = [c_0 \ \cdots \ c_{m-1}]$, construct the approximating curve with the least-squared error using m' control points $c' = [c'_0 \ \cdots \ c'_{m-1}]$, where $m' < m$. Here, we will assume that both curves are endpoint-interpolating uniform B-spline curves.

The multiresolution analysis framework allows this problem to be solved trivially for certain values of m and m'. Assume for the moment that $m = 2^j + 3$ and $m' = 2^{j'} + 3$ for some nonnegative integers $j' < j$. Let c^j be a column matrix containing the control points of c. Then the control points c' of the approximating curve are given by

$$c' = A^{j'+1}A^{j'+2} \cdots A^j c^j$$

In other words, we simply run the decomposition algorithm, as described by equation (7.4), until a curve with just m' control points is reached. Note that this process can be performed at interactive speeds for hundreds of control points using the linear-time LU decomposition algorithm described in Section 7.3.2.

One notable aspect of the multiresolution curve representation is its discrete nature. Thus, it is easy to efficiently construct approximating curves with 4, 5, 7, 11, or any $2^j + 3$ control points, for any integer level j. However, there is no obvious way to quickly construct curves that have "levels" of smoothness in between.

(a)   (b)   (c)

FIGURE 8.1   Smoothing a curve continuously: (a) level 8.0; (b) level 5.4; (c) level 3.1.

One solution that works well in practice is to define a *fractional-level curve* $\gamma^{j+\mu}(t)$ for some $0 \leq \mu \leq 1$ using linear interpolation between its two nearest integer-level curves $\gamma^{j}(t)$ and $\gamma^{j+1}(t)$:

$$\gamma^{j+\mu}(t) = (1 - \mu)\,\gamma^{j}(t) + \mu\,\gamma^{j+1}(t)$$
$$= (1 - \mu)\,\mathbf{\Phi}^{j}(t)\,\mathbf{c}^{j} + \mu\,\mathbf{\Phi}^{j+1}(t)\,\mathbf{c}^{j+1} \qquad (8.1)$$

These fractional-level curves allow smoothing to take place at any continuous level. In an interactive setting, a user can continuously vary a curve from its smoothest form (with four control points) to its most finely detailed version (with all m control points). Some fractional-level curves are shown in Figure 8.1.

8.3   Editing a curve

Suppose we have a curve with control points $\mathbf{c}^{J}$ and all of its low-resolution and detail parts $\mathbf{c}^{0}, \ldots, \mathbf{c}^{J-1}$ and $\mathbf{d}^{0}, \ldots, \mathbf{d}^{J-1}$. Multiresolution analysis allows for two very different kinds of curve editing. If we modify some low-resolution version $\mathbf{c}^{j}$ and then add back in the detail $\mathbf{d}^{j}, \mathbf{d}^{j+1}, \ldots, \mathbf{d}^{J-1}$, we will have changed the overall sweep of the curve (see Figure 8.2, page 113). On the other hand, if we modify the set of detail functions $\mathbf{d}^{j}, \mathbf{d}^{j+1}, \ldots, \mathbf{d}^{J-1}$ but leave the low-resolution versions $\mathbf{c}^{0}, \ldots, \mathbf{c}^{j}$ intact, we will have altered the character of the curve without affecting its overall sweep (see Figure 8.5, page 119). These two types of editing are explored more fully below.

8.3.1   Editing the sweep

Editing the sweep of a curve at an integer level of the wavelet transform is simple. Let $\mathbf{c}^{J}$ be the control points of the original curve $\gamma^{J}(t)$, let $\mathbf{c}^{j}$ be a low-resolution version of $\mathbf{c}^{J}$, and let $\hat{\mathbf{c}}^{j}$

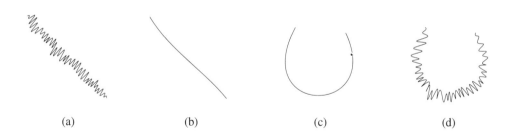

FIGURE 8.2 Changing the overall sweep of a curve without affecting its character. An original curve (a) is reduced to a low-resolution version (b). Once the user modifies the sweep (c), the original detail is reapplied (d).

be an edited version of c^j, given by $\hat{c}^j = c^j + \Delta c^j$. The edited version of the highest-resolution curve $\hat{c}^J = c^J + \Delta c^J$ can be computed through reconstruction:

$$\hat{c}^J = c^J + \Delta c^J$$
$$= c^J + P^J P^{J-1} \cdots P^{j+1} \Delta c^j$$

Note that editing the sweep of the curve at lower levels of smoothing j affects larger portions of the high-resolution curve $\gamma^J(t)$. At the lowest level, when $j = 0$, the entire curve is affected; at the highest level, when $j = J$, only the narrow portion influenced by one original control point is affected. The kind of flexibility that this multiresolution editing allows is suggested in Figures 8.2 and 8.3(a).

In addition to editing at integer levels of resolution, it is natural to ascribe meaning to editing at fractional levels as well. We would like the portion of the curve affected when editing at fractional level $j + \mu$ to interpolate the portions affected at levels j and $j + 1$. Thus, as μ increases from 0 to 1, the portion affected should gradually narrow down from that of level j to that of level $j + 1$, as demonstrated in Figure 8.3(b).

Consider a fractional-level curve $\gamma^{j+\mu}(t)$ given by equation (8.1). Let $c^{j+\mu}$ be the set of control points associated with this curve; that is,

$$\gamma^{j+\mu}(t) = \Phi^{j+1}(t)\, c^{j+\mu} \tag{8.2}$$

Suppose now that one of the control points $c_i^{j+\mu}$ is modified by the user. In order to allow the portion of the curve affected to depend on μ in the manner described above, the system will have to automatically move some of the nearby control points when $c_i^{j+\mu}$ is modified. The distance that each of these control points is moved is inversely proportional to μ: for example,

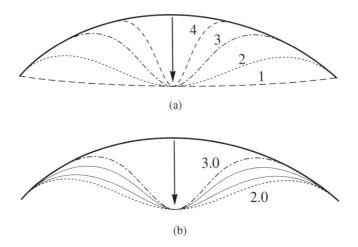

(a)

(b)

FIGURE 8.3   The middle of the dark curve is pulled: (a) integer levels of editing 1, 2, 3, and 4; (b) fractional levels of editing between 2.0 and 3.0.

when μ is near 0, the control points at level $j + \mu$ are moved in conjunction so that the overall effect approaches that of editing a single control point at level j; when $\mu = 1$, the nearby control points are not moved at all, since the modified curve should correspond to moving just a single control point at level $j + 1$.

Let $\Delta c^{j+\mu}$ be a vector describing how each control point of the fractional-level curve is modified: the i-th entry of $\Delta c^{j+\mu}$ is the user's change to the i-th control point; the other entries reflect the computed movements of the other control points. Rather than solving for $\Delta c^{j+\mu}$ explicitly, we will break this vector into two components: a vector Δc^j of changes to the control points at level j and a vector Δd^j of changes to the wavelet coefficients at level j, defined as $\Delta d^j = B^{j+1}\Delta c^{j+1}$. The changes to the high-resolution control points are then reconstructed using a straightforward application of equation (7.6):

$$\Delta c^J = P^J P^{J-1} \cdots P^{j+2} \left(P^{j+1}\Delta c^j + Q^{j+1}\Delta d^j \right) \tag{8.3}$$

We can obtain an expression for $\Delta c^{j+\mu}$ by equating the right-hand sides of equations (8.1) and (8.2) and then applying equation (7.1):

$$\Delta c^{j+\mu} = (1 - \mu)\, P^{j+1}\, \Delta c^j + \mu\, \Delta c^{j+1}$$

Substituting in for Δc^{j+1} using equation (7.6) allows us to write $\Delta c^{j+\mu}$ in terms of Δc^j and Δd^j:

$$\Delta c^{j+\mu} = P^{j+1} \Delta c^j + \mu \, Q^{j+1} \Delta d^j \tag{8.4}$$

Next, define a new vector $\Delta \acute{c}^j$ as the change to control points at level j necessary to make the modified control point $c_i^{j+\mu}$ move to its new position. Also, define $\Delta \acute{c}^{j+\mu}$ to be the user's change to the control points at level $j + \mu$, that is, a vector whose i-th entry is $\Delta c_i^{j+\mu}$ and whose other entries are 0. Note that applying either change alone, $\Delta \acute{c}^j$ or $\Delta \acute{c}^{j+\mu}$, would cause the selected control point to move to its new position; however, the former change would cause a larger portion of the curve to move. In order to have a breadth of change that gradually decreases as μ goes from 0 to 1, we can interpolate between these two vectors, using some interpolation function $g(\mu)$:

$$\Delta c^{j+\mu} = (1 - g(\mu)) \, P^{j+1} \, \Delta \acute{c}^j + g(\mu) \, \Delta \acute{c}^{j+\mu} \tag{8.5}$$

Thus, $\Delta c^{j+\mu}$ will still move the selected control point to its new position, and it will also now control the breadth of change as a function of μ.

Next, if we equate the right-hand sides of equations (8.4) and (8.5) and multiply both sides of the resulting equation by either A^{j+1} or B^{j+1}, we get

$$A^{j+1} \, P^{j+1} \, \Delta c^j + \mu \, A^{j+1} \, Q^{j+1} \, \Delta d^j = (1 - g(\mu)) \, A^{j+1} \, P^{j+1} \, \Delta \acute{c}^j + g(\mu) \, A^{j+1} \, \Delta \acute{c}^{j+\mu}$$
$$B^{j+1} \, P^{j+1} \, \Delta c^j + \mu \, B^{j+1} \, Q^{j+1} \, \Delta d^j = (1 - g(\mu)) \, B^{j+1} \, P^{j+1} \, \Delta \acute{c}^j + g(\mu) \, B^{j+1} \, \Delta \acute{c}^{j+\mu}$$

Finally, applying the "invertibility" identities (equation (7.10) of Table 7.1) allows us to simplify the equations above, giving us the expressions we need:

$$\Delta c^j = (1 - g(\mu)) \, \Delta \acute{c}^j + g(\mu) \, A^{j+1} \, \Delta \acute{c}^{j+\mu} \tag{8.6}$$
$$\Delta d^j = \frac{g(\mu)}{\mu} \, B^{j+1} \, \Delta \acute{c}^{j+\mu}$$

We now have the choice of any function $g(\mu)$ that increases monotonically from 0 to 1, allowing the detail Δd^j to be incorporated gradually. The function $g(\mu) := \mu^2$ is an obvious choice that works well in practice.

The last detail is the definition of $\Delta \acute{c}^j$. We choose the vector that is 0 everywhere, except for one or two entries, depending on the index i of the modified control point. By examining the i-th row of the refinement matrix P^{j+1}, we can determine whether the modified control point is maximally influenced by *one* control point c_k^{j+1} or *two* control points c_k^{j+1} and c_{k+1}^{j+1} at level $j + 1$. In the former case, we define $\Delta \acute{c}_k^j$ as $\Delta c_i^{j+\mu}/P_{i,k}^{j+1}$. In the latter case, we define $\Delta \acute{c}_k^j$ and $\Delta \acute{c}_{k+1}^j$ as $\Delta c_i^{j+\mu}/2P_{i,k}^{j+1}$.

In summary, the modified control points of the highest-resolution curve are computed by running through the steps of this derivation in reverse:

1. Define $\Delta \acute{c}^{j+\mu} := [0, \ldots, 0, \Delta c_i^{j+\mu}, 0, \ldots, 0]^{\mathrm{T}}$.

2. Define $\Delta \acute{c}^j$ and $\Delta \acute{c}^{j+1}$ as described in the previous paragraph.

3. Define Δc^j and Δd^j according to equation (8.6).

4. Construct the offsets to the highest-resolution curve according to equation (8.3).

The fractional-level editing defined here works quite well in practice. Varying the editing level continuously gives a smooth and intuitive kind of change in the region of the curve affected, as suggested by Figure 8.3. Because the algorithmic complexity is only linear in the number of control points, the update is easily performed at interactive rates, even for curves with hundreds of control points.

Editing with direct manipulation

The fractional-level editing described above can be easily extended to accommodate *direct manipulation*, in which the user pulls on a curve directly rather than on its defining control points [2, 39, 42, 62]. In order to directly manipulate a multiresolution curve at level $j + \mu$, we need to know how much the control points at levels j and $j + 1$ should change when the user drags a point of the curve to a new position.

Suppose the user pulls the point $\gamma^{j+\mu}(t_0)$ to a new location $\gamma^{j+\mu}(t_0) + \delta$. We would like to compute a change in the control points satisfying

$$\mathbf{\Phi}^j(t_0) \, \Delta \acute{c}^j = \delta$$
$$\mathbf{\Phi}^{j+1}(t_0) \, \Delta \acute{c}^{j+\mu} = \delta$$

These equations are underdetermined, since the row matrices $\mathbf{\Phi}^j(t_0)$ and $\mathbf{\Phi}^{j+1}(t_0)$ are not invertible. However, we can compute the least-squares change to the control points $\Delta \acute{c}^j$ and $\Delta \acute{c}^{j+\mu}$ by using *pseudo-inverse* matrices [47], which are denoted with a superscript +:

$$\Delta \acute{c}^j = (\mathbf{\Phi}^j(t_0))^+ \, \delta \tag{8.7}$$
$$\Delta \acute{c}^{j+\mu} = (\mathbf{\Phi}^{j+1}(t_0))^+ \, \delta$$

These two equations should be interpreted as applying to each dimension x and y separately. That is, take δ to be a scalar (say, the change in x) and each left-hand side and each pseudo-inverse to be a column matrix of scalars. The modified control points of the highest-resolution curve can then be computed in the same fashion outlined for control-point manipulation by applying equations (8.6) and (8.3).

Note that the first step of the construction, equation (8.7), can be computed in constant time, since for cubic B-splines at most four of the entries of each pseudo-inverse are nonzero. The issue of finding the parameter value t_0 at which the curve passes closest to the user's selection point is a well-studied problem in root finding, which can be handled in a number of ways [106]. In the implementation described by Finkelstein and Salesin, the curve is scan-converted once to find the parameter value and corresponding point closest to the user's selection—an approach that appears to provide a good trade-off between speed and accuracy for an interactive system.

For some applications, it may be more intuitive to drag the highest-resolution curve directly, rather than a smoothed version of the curve. Even when the curve's display resolution is at its highest level, it can be useful to apply a broad change to the curve. This would allow varying levels of detail on the curve to be manipulated by dragging a single point: as the editing resolution is lowered, more and more of the curve is affected. This type of control can be supported quite easily by defining δ as the change in the high-resolution curve at the dragged point $\gamma^J(t_0)$ and using equation (8.7).

Editing a desired portion of a curve

One difficulty with curve manipulation methods is that their effect often depends on the parameterization of the curve, which does not necessarily correspond to the curve's geometric embedding in an intuitive fashion. The manipulation that we have described so far suffers from this same difficulty: dragging at a particular (possibly fractional) level $\ell = j + \mu$ on different points along the curve will not necessarily affect constant-length portions of the curve. However, we can use the multiresolution editing control to compensate for this defect in direct manipulation, as follows.

Let h be a parameter, specified by the user, that describes the desired length of the editable portion of the curve. The parameter h can be specified using any type of physical units, such as screen pixels, inches, or percentage of the overall curve length. The system computes an appropriate editing level ℓ that will affect a portion of the curve of about h units in length, centered at the point $\gamma^J(t_0)$ being dragged.

We estimate ℓ as follows. For each integer-level editing resolution j, let $h^j(t_0)$ denote the length of $\gamma^J(t)$ affected by editing the curve at the point $\gamma^J(t_0)$. The length $h^j(t_0)$ is easily estimated by scan-converting the curve $\gamma^J(t)$ to determine the approximate lengths of its polynomial segments and then summing over the lengths of the segments affected when editing the curve at level j and parameter position t_0. Next, we define j_- as the largest value of j with

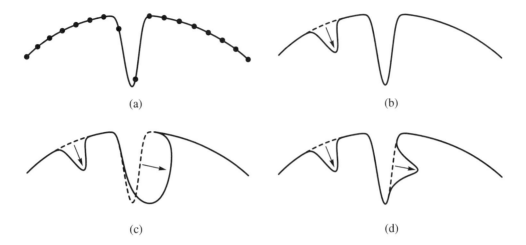

FIGURE 8.4   (a) A curve with a parameterization that changes most rapidly in the middle. Direct manipulation on the left part of the curve (b) affects a much smaller fraction of the curve than it does in the middle (c). A multiresolution curve editor can automatically determine the appropriate editing level needed to affect a specified fraction of the curve (d).

$h^{j_-}(t_0) \geq h$ and j_+ as the smallest value of j with $h^{j_+}(t_0) \leq h$. We use linear interpolation to find a suitable fractional editing level ℓ that lies between j_- and j_+:

$$\ell := \frac{h - h^{j_+}}{h^{j_-} - h^{j_+}} j_- + \frac{h^{j_-} - h}{h^{j_-} - h^{j_+}} j_+$$

Finally, by representing ℓ in terms of an integer level j and fractional offset μ, we can again apply equation (8.7), followed by equations (8.6) and (8.3), as before. Though in general this construction does not *precisely* cover the desired portion h, in practice it yields an intuitive and meaningful control. Figure 8.4 demonstrates this type of editing for a curve with an extremely nonuniform geometric embedding.

In summary, direct manipulation and the kind of multiresolution editing described in this section work well together. With multiresolution direct manipulation, the user can drag the curve directly and effect any range of narrow to broad types of changes, depending on the editing resolution selected. In addition, this sort of multiresolution editing can be used to compensate for any peculiarities of the geometric embedding of the parametric curve, allowing for a much more explicit control over the geometric portion of the curve affected.

FIGURE 8.5   Changing the character of a curve without affecting its sweep.

8.3.2   Editing the character of a curve

Another form of editing that is naturally supported by multiresolution curves is editing the character of a curve without affecting its overall sweep. Let c^J be the control points of a curve, and let $c^0, \ldots, c^{J-1}, d^0, \ldots, d^{J-1}$ denote the components of its multiresolution decomposition. Editing the character of the curve is simply a matter of replacing the existing set of detail functions $d^j, \ldots, d^{J-1}$ with some new set $\hat{d}^j, \ldots, \hat{d}^{J-1}$ and reconstructing.

With this approach, it is possible to develop a "curve character library" that contains different detail functions, which can be applied interchangeably to any set of curves. The detail functions in the library can be extracted from hand-drawn strokes; procedural methods of generating detail functions are also possible. Figure 8.5 demonstrates how the character of curves in an illustration can be modified with various detail styles. The interactive illustration system used to create this figure is described by Salisbury et al. [105].

8.3.3   Orientation of detail

A parametric curve in two dimensions is most naturally represented as two separate functions, one in x and one in y: $\gamma(t) = (\gamma_x(t), \gamma_y(t))$. Thus, it seems reasonable to represent both the control points c^j and detail functions d^j using matrices with separate columns for x and y. However, encoding the detail functions in this manner embeds all of the detail of the curve in a particular x-y orientation. As Figure 8.6 demonstrates, this representation does not always provide the most intuitive control when the sweep of a curve is edited.

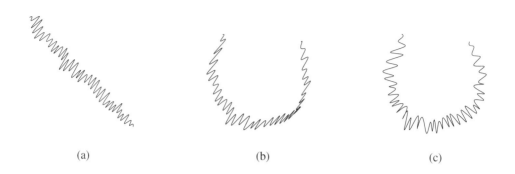

(a)   (b)   (c)

FIGURE 8.6 Editing the sweep of a curve (a) using detail oriented with respect to fixed x- and y-axes (b), and using detail oriented relative to the tangent of the curve (c).

As an alternative, we employ a method similar to one that Forsey and Bartels [39] suggested for hierarchical B-splines: Each detail coefficient d_i^j specifies a change relative to the tangent and normal directions of the lower-resolution curve $\gamma^{j-1}(t)$. The tangent and normal are computed at the parameter value t_0 corresponding to the peak of the wavelet $\psi_i^j(t)$. Note that the curve $\gamma(t)$ is no longer a simple linear combination of the scaling functions $\mathbf{\Phi}^0$ and wavelets $\mathbf{\Psi}^j$; instead, reconstructing the curve requires that a change of coordinates be applied to the wavelet coefficients $\boldsymbol{d}^j$ at each level of reconstruction. However, this process is linear in the number of control points, so it does not increase the computational complexity of the algorithm.

It is possible to use either normalized or unnormalized versions of the tangent–normal reference frame; the two alternative versions yield different but equally reasonable behavior. Figure 8.5 uses the unnormalized tangents, while the rest of the figures in this chapter use normalized tangents.

8.4   Scan conversion and curve compression

When we use curve character libraries and other multiresolution editing features, it is easy to create very complex curves with hundreds or thousands of control points. In many cases (as in this book), these curves are printed in a very small form. Conventional scan conversion methods that use all the complexity of these curves are wasteful, both in terms of the network traffic caused by sending such large files to the printer and in terms of the processing time required by the printer to render curves with many control points within a few square pixels. We therefore explore a form of curve compression that is suitable for the purposes of scan conversion. The algorithm we describe finds a curve that approximates the original curve within a specified error tolerance. However, it does not require any particular continuity constraints, as are usually required in data-fitting applications.

In Section 8.2 we showed how applying a single step of a spline-wavelet filter bank produced an approximating curve with half as many control points as the original curve. This type of smoothing can be used to find approximations that minimize a least-squares (L^2) measure of error. However, an L^2 metric is not very useful for scan conversion because an approximate curve $\hat{\gamma}(t)$ can be arbitrarily far from the original curve $\gamma^J(t)$ and still achieve a particular L^2 error bound, as long as it deviates from the original over a small enough range of parameter values. In order to scan-convert a curve within some guaranteed precision—measured, say, in terms of maximum deviation in printer pixels—we need to bound the L^∞ norm of the error. There are many ways to achieve such a bound. The method described here is simple and fast, and methods that achieve higher compression ratios are certainly possible.

Let s_i^j (with $0 \le i \le 2^j - 1$) be a segment of the cubic B-spline curve $\gamma^j(t)$, defined by the four control points $c_i^j, \ldots, c_{i+3}^j$. Note that each segment s_i^j corresponds to exactly two (more detailed) segments s_{2i}^{j+1} and s_{2i+1}^{j+1} at level $j + 1$. Our objective is to build a new approximating curve $\hat{\gamma}(t)$ for $\gamma(t)$ by choosing different segments at different levels such that $\| \hat{\gamma}(t) - \gamma^J(t) \|_\infty$ is less than some user-specified tolerance ε for all values of t.

Assume, for the moment, that we have some function $ErrBound(s_i^j)$ that returns a bound on the L^∞ error incurred from using the segment s_i^j of some approximate curve $\gamma^j(t)$ in place of the original segments of $\gamma^J(t)$ to which it corresponds. We can scan-convert a curve to within any error tolerance ε by passing to the recursive routine $DrawSegment$ the single segment s_0^0 corresponding to the lowest-level curve $\gamma^0(t)$. This routine recursively divides the segment to varying levels so that the collection of segments it produces approximates the curve to within ε.

> **procedure** $DrawSegment(s_i^j)$
> > **if** $ErrBound(s_i^j) < \varepsilon$ **then**
> > > Output segment s_i^j as a portion of $\hat{\gamma}(t)$
> > **else**
> > > $DrawSegment(s_{2i}^{j+1})$
> > > $DrawSegment(s_{2i+1}^{j+1})$
> > **end if**
> **end procedure**

To construct the $ErrBound$ function, let $\boldsymbol{M}^j$ be the B-spline–to–Bézier-basis conversion matrix [3] for curves with $2^j + 3$ control points, and let $\boldsymbol{e}^j$ be a column vector with entries e_i^j defined by

$$e^j := M^j \, Q^j \, d^{j-1}$$

The vector e^j provides a measure of the distance that the Bézier control points migrate when reconstructing the more detailed curve at level j from the approximate curve at level $j - 1$. Since Bézier curves are contained within the convex hull of their control points, the magnitude of entries e^j provide conservative bounds on approximations to the curve due to truncating wavelet coefficients. (Bézier control points are used to give tighter bounds than B-spline control points.)

A bound δ_i^j on the L^∞ error incurred by replacing segment s_i^j with its approximation at level $j - 1$ is given by

$$\delta_i^j \leq \max_{i \leq k \leq i+3} \left\{ \| e_k^j \|_2 \right\}$$

The *ErrBound* routine can then be described recursively as follows:

function *ErrBound*(s_i^j)
   if $j = J$ **then**
      return 0
   else
      return $\max \{ ErrBound(s_{2i}^{j+1}) + \delta_{2i}^{j+1}, ErrBound(s_{2i+1}^{j+1}) + \delta_{2i+1}^{j+1} \}$
   end if
end function

An efficient implementation of the *ErrBound* routine would use dynamic programming or an iterative (rather than recursive) function to avoid recomputing error bounds. In practice, the routine is fast enough in its recursive form that this optimization does not appear to be necessary, at least for scan-converting curves with hundreds of control points.

The approximate curve $\hat{\gamma}(t)$ is described by a set of Bézier segments, which we use to generate a PostScript file. Note that the scan conversion algorithm, as described, produces approximate curves $\hat{\gamma}(t)$ that are not even continuous where two segments of different levels abut. Since we are only concerned with the absolute error in the final set of pixels produced, relaxing the continuity of the original curve is reasonable for scan conversion. We can achieve positional continuity, however, without increasing the prescribed error tolerance, by simply averaging together the end control points for adjacent Bézier segments as a postprocess. Curves made continuous in this manner look slightly better than their discontinuous counterparts; they also have a more compact representation in PostScript. Figure 8.7 demonstrates compression of the same curve rendered at different sizes.

(a)

(b)

FIGURE 8.7 Scan-converting a curve within a guaranteed maximum error tolerance. (a) From left to right, the figures used 5%, 21%, 46%, and 78% of the possible number of Bézier segments. Error is less than 1/300 inch. (b) The same curves, drawn at constant size.

MULTIRESOLUTION TILING

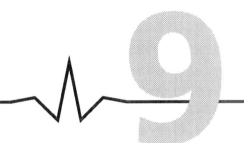

1. Previous solutions to the tiling problem — 2. The multiresolution tiling algorithm — 3. Time complexity — 4. Tiling examples

Reconstructing a surface from a set of planar contours is an important problem in medical imaging, biological research, and geological research. The problem can be reduced to several subproblems [84, 85], one of which, the *tiling problem*, is the subject of this chapter. The objective of the tiling problem is to construct a polyhedron from two planar polygons. The two polygons form two of the faces of the polyhedron, while each remaining face of the polyhedron is a triangle joining an edge of one contour to a vertex of the other contour. As an example, Figure 9.1 shows a pair of contours along with a solution to the tiling problem.

The difficulty of the tiling problem lies in constructing the "best" surface between two contours. In this chapter we will describe how wavelets can be used to efficiently construct a tiling that approximates an optimal solution. This approach, which we refer to as the *multiresolution tiling algorithm*, was first described in detail by Meyers [83, 84]. The work presented by Meyers makes extensive use of single-knot wavelets very similar to the ones developed in Section 7.4.3 to hierarchically decompose polygonal contours.

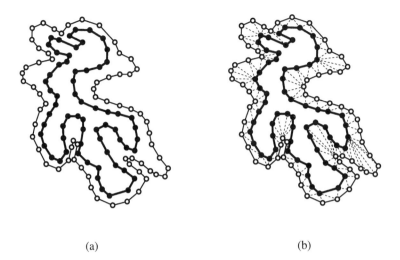

(a)   (b)

FIGURE 9.1   (a) A pair of planar polygon contours; (b) a solution to the tiling problem for these contours.

9.1   Previous solutions to the tiling problem

Many algorithms have been devised to solve the tiling problem. A method that computes a tiling that is optimal with respect to a certain goal function was devised by Keppel [67] and later improved by Fuchs et al. [43]. We will refer to this latter algorithm as the *optimizing algorithm*.

The optimizing algorithm begins by reducing the tiling problem to a graph-searching problem. As illustrated in Figure 9.2, a graph can be constructed from a pair of contours by first associating a node with each of the edges that could connect one contour to the other (these edges are called *spans*). Next, we consider each pair of nodes and connect them with an arc if the two spans they represent share a vertex and the other two vertices are adjacent on a contour. These conditions guarantee that each arc corresponds to a triangle formed by connecting a segment on one contour to a vertex on the opposite contour. By convention, the arcs are directed so that the corresponding triangle's contour edge is traversed in a counterclockwise direction.

The nodes and arcs that result from this construction form a toroidal graph. Tilings are represented by cycles that visit each row and each column of the graph. The optimizing tiling algorithm associates a cost with each arc (for example, the area of the triangle represented by the arc), thereby reducing the tiling problem to one of finding the least-cost cycle in the graph. For the sake of brevity, we will not explain how to find an optimal cycle; suffice it to say it is a

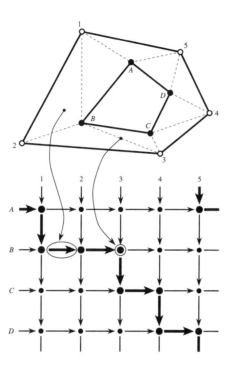

FIGURE 9.2 The optimizing tiling algorithm maps spans (dashed lines) of a tiling to nodes of a toroidal graph and maps triangles of the tiling to edges in the graph. The tiling shown is represented by the cycle formed by the bold edges. Other tilings can be constructed from other cycles in the graph.

well-studied problem in discrete optimization that can be solved using dynamic programming techniques [43].

Note that although tilings produced by the optimizing algorithm are "optimal" with respect to a quantitative goal function, they are not necessarily optimal according to a user. For example, consider the two contours in Figure 9.3(a), representing anatomical data from the human brain. Each of the contours has several indentations along its perimeter that represent the invaginations or sulci of the cerebral cortex. In a correct reconstruction, these indentations should be matched so that the sulci remain continuous between the two layers. However, the tiling produced by the optimizing algorithm, shown in Figure 9.3(b), fails to match the two indentations closest to the bottom of the figure. Thus, it may be possible to improve the results of the optimizing algorithm by allowing some degree of user interaction.

The major drawback of the optimizing algorithm is its computational expense: for a pair of contours containing a total of n vertices, the optimizing algorithm requires $O(n^2 \log n)$ time and $O(n^2)$ space to construct a tiling. In particular, the time required to reconstruct a surface

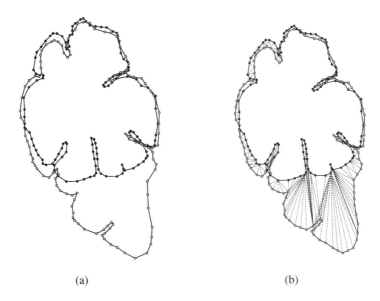

(a)   (b)

FIGURE 9.3   (a) A pair of contours obtained from the cerebral cortex of the human brain. The two contours consist of 128 vertices (closed dots) and 114 vertices (open dots). (b) An unacceptable tiling produced by the optimizing algorithm, demonstrating the need for user interaction.

becomes prohibitively expensive for interactive applications as the contours become large (contours containing thousands of vertices are not uncommon in actual data sets). We therefore turn to faster, nonoptimizing methods.

In order to achieve interactive speeds, a number of algorithms have been proposed that attempt to find a reasonable tiling without using global optimization. Of these nonoptimizing methods (discussed further by Meyers et al. [85]), we will mention just two algorithms that construct tilings in linear time: the "greedy" methods of Ganapathy and Dennehy [45] and Christiansen and Sederberg [16]. Both methods construct a tiling beginning with an edge assumed to be a good starting point. They then sequentially advance along one of the contours, connecting the current vertex on one contour to the next vertex on the other. The Christiansen-Sederberg algorithm attempts to minimize the sum of edge lengths by always selecting the shorter of the two possible edges at each step. The Ganapathy-Dennehy algorithm always selects the edge that minimizes the difference between the fractions of arc length traversed along each contour. The results of these algorithms applied to the human brain contours are shown in Figure 9.4. Each of the algorithms gets "confused" by a local configuration that is not well modeled by its heuristic. Though faster to compute, the resulting tilings are significantly worse than the one produced by the optimizing algorithm.

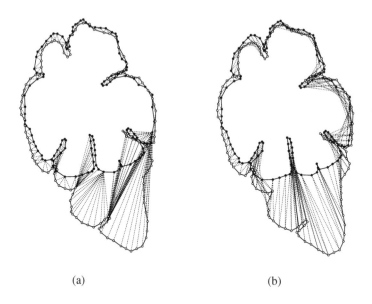

(a)   (b)

FIGURE 9.4   Tilings of the contours in Figure 9.3 produced by linear-time methods: (a) Christiansen and Sederberg's algorithm; (b) Ganapathy and Dennehy's algorithm.

9.2   The multiresolution tiling algorithm

Multiresolution analysis motivates a new approach to solving the tiling problem. The key steps in the multiresolution tiling algorithm are summarized below and illustrated in Figure 9.5.

1. Reduce the size of the problem by using a wavelet decomposition to find low-resolution approximations to the original contours.

2. Use the optimizing algorithm to produce a tiling of the low-resolution contours.

3. Perform one step of reconstruction to incorporate one or several wavelet coefficients into each of the contours, thereby adding vertices and edges to the current tiling.

4. Improve the tiling by applying local optimizations to edges in the neighborhood of newly added edges.

5. Repeat steps 3 and 4 until all of the original vertices are present.

We will describe each of these steps in more detail in the sections that follow.

Note that a tiling produced by the multiresolution algorithm is not guaranteed to be globally optimal with respect to the goal function used to compute the low-resolution tiling. However, since the algorithm begins with an optimized base case and maintains local optimality, the final tiling is often very nearly identical to that computed by the optimizing algorithm. Significant differences between the methods occur most often in areas where the contours' shapes are very different. In such situations, it is often the case that neither method produces an acceptable result and user intervention is required (as with the contours in Figure 9.3). The speed of the multiresolution algorithm then gives it a clear advantage over the optimizing algorithm.

9.2.1   Contour decomposition

The multiresolution tiling algorithm begins by decomposing each input contour into a set of wavelet coefficients and a lower-resolution contour. For this process, we use the single-knot wavelet filter-bank method described in Section 7.4.3. Recall that the filter bank requires as input a sequence of coefficients whose length is a power of two. If the number of vertices in a contour is not a power of two, we add vertices by iteratively splitting the longest edge and adding a new vertex at its midpoint. We use a priority queue to efficiently maintain a list of edges sorted by length.

9.2.2   Optimizing the base case tiling

The next step of the multiresolution tiling algorithm takes the pair of low-resolution contours, called the *base case*, and applies the optimizing algorithm to produce a tiling. The size of the base case influences both the speed with which a tiling is computed and the quality of the final result.

Reducing the problem to a small base case allows the computer to execute the optimizing algorithm quickly; however, the quality of the resulting tiling may suffer if the low-resolution contours do not adequately approximate the original contour shapes. Using a large base case requires more time, but may produce a better result. One option is to have the user to specify the base case size, allowing the user to establish the trade-off between acceptable tiling result and execution time. Meyers [83] reports that in the absence of a user-specified base case size, 64 vertices is usually a reasonable choice.

9.2.3   Reconstruction

Once a low-resolution tiling is obtained, we need to reinsert some of the detail that was removed from the contours. The reconstruction of a contour from its low-resolution approximation can be done in two different ways. One method, called *filter-bank reconstruction*, rein-

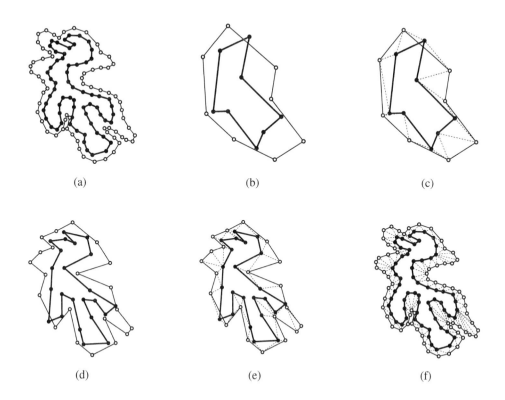

FIGURE 9.5 The main steps of the multiresolution tiling algorithm: The input contours (a) are reduced to low-resolution versions (b), and tiled by the optimizing algorithm (c). Wavelets are reinserted to produce intermediate-resolution contours (d) and local optimization is performed (e). This process is repeated until all vertices are present (f).

serts the wavelets one filter-bank level at a time. The other method, called *single-wavelet reconstruction*, reinserts wavelets one at a time, from largest to smallest. Filter-bank reconstruction requires only $O(n)$ time, while single-wavelet reconstruction requires $O(n \log n)$. However, single-wavelet reconstruction also allows for contour compression and appears to produce somewhat better results. We discuss both methods in more detail below.

Filter-bank reconstruction

With filter-bank reconstruction, wavelet coefficients are returned to the contours one filter-bank level at a time, from the coarsest level to the finest, according to the following recipe:

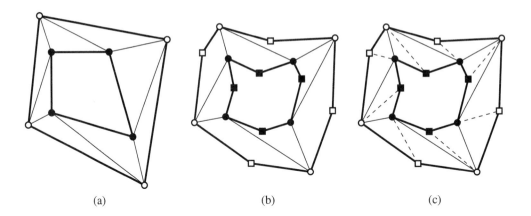

         (a)   (b)   (c)

FIGURE 9.6   Filter-bank reconstruction: (a) a tiling of two low-resolution contours. (b) One level of filter-bank reconstruction adds the vertices represented by squares. (c) The dashed edges are added to the tiling.

1. Place all the edges in the current tiling that connect one contour to the other in a list of *suspect edges*.

2. Construct a new polygon for each contour using one level of the filter bank. This step splits each edge of both contours and, in general, moves every vertex. As a result, each triangle in the tiling becomes a quadrilateral with three vertices on one of the contours and the fourth on the opposite contour, as shown in Figure 9.6(a) and (b).

3. For each new vertex added to a contour, construct an edge from that vertex to the quadrilateral vertex on the other contour, splitting the quadrilateral into two triangles, as shown in Figure 9.6(c).

4. Use the list of suspect edges to perform local optimizations on the tiling as described in Section 9.2.4.

5. Repeat steps 1 through 4 until the resolution of the input contours is reached. There will be $n - m$ iterations for input contours of 2^n vertices and lowest-resolution contours of 2^m vertices.

Single-wavelet reconstruction

The filter-bank reconstruction process doubles the resolution of each contour at each step, and requires that wavelet coefficients be added in the opposite of the order in which they were computed during the decomposition step. Alternatively, by adding wavelet coefficients one at a time, it is possible to use them in any desired order, regardless of the resolution level from which they were obtained. It is particularly useful to reconstruct by adding the wavelet coefficients in order of decreasing magnitude.

Adding wavelets in decreasing order has two benefits. First, it provides a form of compression: by using only the wavelet coefficients with a magnitude larger than some threshold value, we reduce the number of vertices in a contour while preserving as much detailed structure as possible. Second, reconstructing by adding the largest wavelets first causes the contours to approach their original shape as rapidly as possible. Intuitively, it seems plausible that a better tiling should result because the local optimization should operate on a closer approximation to the final shape earlier in the process. Indeed, in practice, single-wavelet reconstruction seems to produce better tilings than the filter-bank method.

The single-wavelet method begins with the same decomposition of contours as the filter-bank method. Then, starting from an optimized low-resolution tiling, the single-wavelet method proceeds as follows:

1. Select a wavelet to add by alternating between contours in each iteration and choosing the remaining wavelet in the appropriate contour with the largest x and y coefficients.

2. Multiply the chosen x and y wavelet coefficients by the corresponding wavelet basis function and add the result to the contour. Figure 9.7 shows how two piecewise-linear functions in one dimension are added once the two sets of knots are merged.

3. For each vertex that is added to a contour, create an edge connecting it to the appropriate vertex of the opposite contour (just as in step 3 of the filter-bank reconstruction method).

4. Place all the edges incident on a newly added or moved vertex in a list of suspect edges.

5. Use the list of suspect edges to perform local optimizations on the tiling as described in Section 9.2.4.

6. Repeat steps 1 through 5 until all wavelets have been incorporated or until the magnitudes of the remaining wavelet coefficients are below a threshold value.

Step 2 of this process is illustrated in Figure 9.7 for two cases of the addition of a single wavelet to a one-dimensional piecewise-linear function. The process is the same for a two-dimensional contour, except that the x- and y-coordinates of a vertex are modified by the x and y components of a wavelet coefficient.

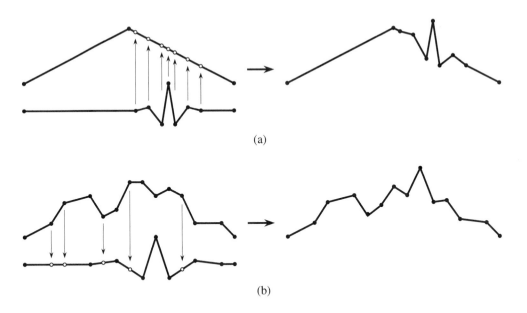

(a)

(b)

FIGURE 9.7   Illustration of single-wavelet reconstruction in one dimension: (a) a high-resolution wavelet is added to a low-resolution function (open dots represent knots in the wavelet that are added to the function); (b) a low-resolution wavelet is added to a high-resolution function (open dots represent knots in the function that are added to the wavelet).

9.2.4   Local optimization

In both the filter-bank approach and the single-wavelet approach to reconstruction, a list of suspect edges is created as the contours are modified by the addition of new wavelets. The local optimization step examines each of these suspect edges to determine whether an *edge swap* is possible. As Figure 9.8 illustrates, an edge can be swapped when it is shared by a triangle containing an edge of one contour and a triangle containing an edge of the opposite contour.

If it is possible to swap a suspect edge, the local optimization next determines whether a swap of its orientation reduces the goal function that is being minimized. If so, the edge is swapped, and both edges adjacent to the swapped edge are added to the list of suspect edges. The local optimization process terminates when this list is empty.

FIGURE 9.8   Edge swapping: the dashed edge can be swapped.

9.3   Time complexity

The asymptotic time complexity of a tiling algorithm determines whether it can be used interactively for large contours. As we have already remarked, the optimizing algorithm takes $O(n^2 \log n)$ time for a pair of contours with n vertices.

The first step of the multiresolution tiling algorithm adds vertices to ensure that the number of vertices in each contour is a power of two. This process does not affect the complexity of computing a tiling, since the number of vertices in a contour is at most doubled. With the appropriate choice of a priority queue implementation, adding vertices requires at most $O(n \log n)$ time for a contour with n vertices. Once each contour's size is a power of two, the contours can be decomposed using the filter-bank method in $O(n)$ time. The base case tiling is then constructed in constant time, as the lowest resolution is independent of the size of the input.

Reconstruction of the original tilings utilizes either the filter-bank or the single-wavelet method. Disregarding any local optimizations of the tiling that occur during reconstruction, filter-bank reconstruction can be accomplished in $O(n)$ time, while the single-wavelet approach takes $O(n \log n)$ time. Since the asymptotic complexity of the step that adds vertices to the original contours is already $O(n \log n)$, the benefits associated with adding wavelets in sorted order typically outweigh the cost of sorting the coefficients and reconstructing one wavelet at a time.

Local optimization occurs each time a wavelet or group of wavelets introduces a new vertex into a contour. The complexity of this step is more difficult to analyze. One can imagine a situation in which a single edge swap causes a cascade of further edge swaps that take $O(n)$ time. However, Meyers gives empirical data demonstrating that for each vertex added during reconstruction, the number of edges examined during local optimization is very nearly constant for contours ranging in size from 128 to 1024 vertices [83]. As a result, the time complexity of reconstruction is not increased by the local optimization that occurs in each iteration.

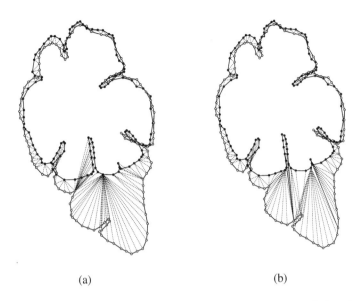

FIGURE 9.9   (a) A tiling of the contours in Figure 9.3 produced by the multiresolution algorithm; (b) a tiling produced by the optimizing algorithm.

9.4   Tiling examples

As mentioned earlier, the contours shown in Figure 9.3 present a difficult instance of the tiling problem. A trained anatomist would recognize that each of the seven indentations on one contour should be linked to a corresponding indentation on the other contour by edges at its inner extremum. The multiresolution tiling algorithm's solution to this problem is shown in Figure 9.9(a). This tiling compares favorably to the optimizing algorithm's solution, which appears in Figure 9.9(b). Note that there are areas in both tilings that may not be acceptable according to the criterion that the indentations be linked. In general, the "correct" tiling is ambiguous and depends on the nature of the material from which the contours are derived. In this case at least, the multiresolution algorithm connects six out of seven indentations, while the optimizing algorithm connects only five.

In addition, the single-wavelet multiresolution approach can easily incorporate data compression by leaving out the wavelet coefficients of smallest magnitude. Figure 9.10 shows a sequence of tilings reconstructed using different compression thresholds. As the threshold increases, the number of vertices in the contours decreases significantly, while the overall shapes of the contours retain much of the original detail. The least detailed tiling, shown in

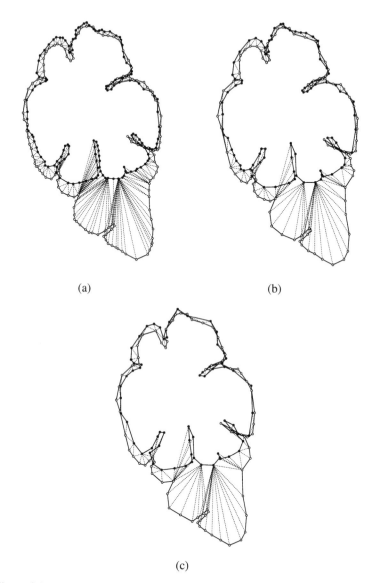

(a)　　　　　　　　　　　　(b)

(c)

FIGURE 9.10　Tilings of the contours in Figure 9.3 computed by the single-wavelet multiresolution algorithm with a compression threshold. When multiplied by the magnitude of the largest wavelet coefficient, the threshold determines the magnitude of the smallest coefficient used. (a) Threshold factor 0.001; (b) threshold factor 0.0025; (c) threshold factor 0.005.

Figure 9.10(c), may be adequate for many purposes. This low-resolution solution requires significantly less space to store and less time to display than the fully detailed tiling.

III

SURFACES

SURFACE WAVELETS

1. Overview of multiresolution analysis for surfaces — 2. Subdivision surfaces —
3. Selecting an inner product — 4. A biorthogonal surface wavelet construction —
5. Multiresolution representations of surfaces

Surfaces play a central role in many three-dimensional computer graphics applications. Objects with flat faces are naturally represented by polyhedral meshes, while objects with curved surfaces are typically represented by tensor-product or triangular spline patches. We would like to construct hierarchical representations of all of these types of objects in order to provide the opportunity for compression, multiresolution editing, and many of the other operations that we've seen applied to images and curves.

A tensor-product surface patch is one type of surface representation that is quite easily converted into a multiresolution form. In 1988, Forsey and Bartels [41] described a hierarchical framework for tensor-product surface constructions called *hierarchical B-splines*. Their framework creates an *overrepresentation* of the geometry—in other words, there may be more than one way of representing a given tensor-product surface as a hierarchical B-spline.

Alternatively, a wavelet representation for tensor-product B-spline surfaces can be constructed by applying either the standard or nonstandard tensor-product construction, described in Chapter 3, to the spline wavelets of Section 7.3.2. A wavelet basis such as this provides a unique representation for every tensor-product surface, and it requires the same amount of storage as the surface's original control points. The wavelet basis allows us to perform on surfaces many of the operations described in Chapter 8 for curves. For instance, in Color Plate 8 a tensor-product cubic B-spline patch is shown after it has been edited at various levels of detail.

Unfortunately, tensor-product constructions are limited in the kinds of shapes they can model seamlessly. In particular, tensor products can only be used for functions parameterized on $\mathbb{R}^2$. They are not applicable to functions defined on more general topological domains, such as the spherical domains shown (Color Plates 9 and 12).

In this chapter we show how multiresolution analysis can be extended to functions defined on two-dimensional domains of arbitrary topological type. (The topological type of a two-dimensional surface or domain refers to its boundary curves as well as its genus.) The extension of multiresolution analysis to arbitrary topological domains is based on creating refinable scaling functions using recursive surface subdivision, in complete analogy with our development of multiresolution analysis for curves.

We begin this chapter with an overview of multiresolution analysis for surfaces. Then, in Section 10.2, we describe the basics of subdivision surfaces and show that subdivision induces nested spaces and refinable scaling functions. In the rest of the chapter, we present a construction of biorthogonal wavelets for surfaces that are similar in spirit to the single-knot wavelets presented in Section 7.4.3.

The generalization of multiresolution analysis to arbitrary topological domains significantly broadens the class of applications to which multiresolution analysis can be applied. Chapter 11 is devoted to an exploration of several such applications, including surface compression, continuous level-of-detail control, progressive transmission across a low-bandwidth network, and multiresolution editing of surfaces.

10.1 Overview of multiresolution analysis for surfaces

Although the mathematics in the remainder of this chapter is somewhat involved, the resulting algorithms are relatively simple. Before diving into the details, we present here a brief overview of how the method can be applied to decompose the polyhedral object shown in Figure 10.1(a).

The idea behind multiresolution analysis for surfaces is the same as it was for images and curves: we split a high-resolution surface (in this case a polyhedral bust of Spock) into a low-resolution part and a detail part. For example, the low-resolution part of the polyhedron in Figure 10.1(a) is shown in Figure 10.1(b). The vertex positions in 10.1(b) are computed as weighted averages of the vertex positions in 10.1(a). Again, just as it was for curves, the computation can be expressed as multiplication by a matrix A^j. The wavelet coefficients repre-

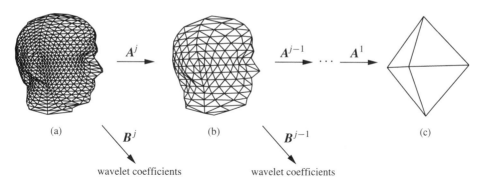

FIGURE 10.1   Decomposition of a polyhedral surface.

senting the detail can similarly be computed by multiplying by a matrix B^j. This process is recursively applied to the low-resolution part until the coarsest representation of the surface is obtained, as shown in Figure 10.1(c).

The analysis filters A^j and B^j can be inverted as in equation (7.10) to produce synthesis filters P^j and Q^j. Synthesis can be viewed more concretely as involving two steps: splitting each triangular face of the low-resolution polyhedron into four subtriangles by introducing new vertices at edge midpoints and perturbing the resulting collection of vertices according to the wavelet coefficients.

The difficulty in creating a multiresolution analysis for surfaces is in designing the four analysis and synthesis filters so that

- the low-resolution versions are good approximations to the original object

- the magnitude of a wavelet coefficient provides some measure of the error introduced when that coefficient is set to zero

- analysis and synthesis have time complexities that are linear in the number of vertices

10.2   Subdivision surfaces

In this section, we show that recursive subdivision applied to surfaces leads to a collection of refinable scaling functions and hence to a sequence of nested linear spaces, as required by multiresolution analysis. In complete analogy with curves, only surfaces generated through subdivision can be hierarchically decomposed using multiresolution analysis.

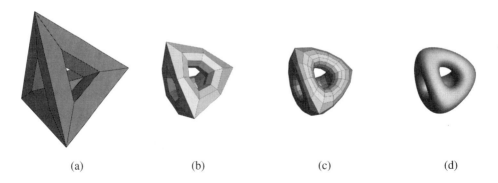

(a)   (b)   (c)   (d)

FIGURE 10.2   Catmull-Clark subdivision: (a) the control mesh M^0; (b) the mesh M^1; (c) the mesh M^2; (d) the limit surface σ. Notice that each quadrilateral face is split into four subfaces.

Chaikin's use of subdivision to create curves inspired Catmull and Clark [7] and simultaneously Doo and Sabin [29, 30] to use subdivision to create surfaces. This work represented a major breakthrough in surface modeling, as it provided the first method of constructing smooth surfaces of arbitrary topological type. Although subdivision surfaces received little attention from the computer graphics community for many years, they are now the subject of new interest. This renewed interest is based in part on the intimate connection between subdivision and multiresolution analysis.

Whereas a subdivision curve is created by iteratively refining a control polygon, a subdivision surface is created by iteratively refining a *control polyhedron* (also called a *control mesh*) M^0 to produce a sequence of increasingly faceted meshes $M^1, M^2, \ldots$ that converge to a surface

$$\sigma := \lim_{j \to \infty} M^j$$

Each refinement step consists of splitting and averaging substeps, in analogy to the steps for curves. However, there are two very different types of splitting steps: *face schemes*, which split faces, and *vertex schemes*, which (not surprisingly) split vertices. Whereas Catmull and Clark developed a face scheme that generalizes tensor-product cubic B-splines (Figure 10.2), Doo and Sabin developed a vertex scheme that generalizes tensor-product quadratic B-splines (Figure 10.3).

In this text, we are going to focus our attention on face schemes, which can be further classified as being either *triangular*, where the meshes have triangular faces, or *quadrilateral*,

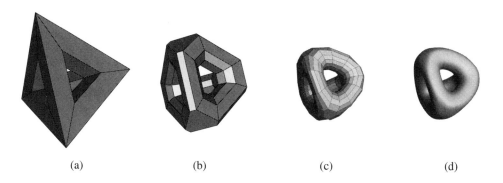

(a)   (b)   (c)   (d)

FIGURE 10.3 Doo-Sabin subdivision: (a) the control mesh M^0; (b) the mesh M^1; (c) the mesh M^2; (d) the limit surface σ. In this scheme, a vertex surrounded by n faces is split into n subvertices, one for each face.

where mesh faces are bounded by four vertices and four edges (but the vertices need not be co-planar). As indicated in Figure 10.2, the Catmull-Clark scheme is quadrilateral. Triangular schemes are somewhat simpler to describe and implement, so we will focus the rest of our discussion on them.

For triangular subdivision, the *splitting step* breaks each face of M^{j-1} into four subfaces by introducing edge midpoints to create a new mesh $\mathring{M}^j$, as illustrated in Figure 10.4. The *averaging step* then perturbs the vertices of $\mathring{M}^j$ to create a mesh M^j that is structurally identical to $\mathring{M}^j$ (that is, M^j and $\mathring{M}^j$ have the same number of vertices, edges, and faces and are connected together in the same way). The position of each vertex of M^j is given by a weighted average of the positions of nearby vertices of $\mathring{M}^j$.

> **Example:** The simplest subdivision scheme is *polyhedral subdivision*, in which the averaging step leaves all vertex positions unchanged. The limit surface of polyhedral subdivision is therefore the same as the control polyhedron M^0. This surface scheme is entirely analogous to the piecewise-linear curve scheme described in Chapter 6, which splits each line segment at its midpoint and leaves the resulting vertices unchanged. ∎

> **Example:** *Loop's scheme* [73] is the simplest triangular subdivision scheme that creates smooth surfaces. Figure 10.5 illustrates how a control mesh is repeatedly subdivided to reach a smooth surface in the limit. Rather than generalizing tensor-product constructions, as do the Catmull-Clark and Doo-Sabin schemes, Loop's scheme generalizes a particular triangular B-spline (the three-direction quartic box spline).

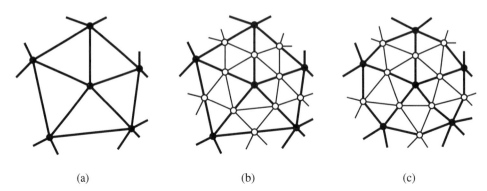

FIGURE 10.4   The splitting step for triangular face schemes: (a) the situation around a vertex of M^{j-1}; (b) the new mesh $\overset{\circ}{M}{}^{j}$ created by the splitting step; (c) the mesh M^{j} created by the averaging step.

To describe the averaging step in Loop's scheme, we will use the following notation: let v be a vertex of M^{j} associated with the vertex $\overset{\circ}{v}$ of $\overset{\circ}{M}{}^{j}$, and let $\overset{\circ}{v}_{1}, \ldots, \overset{\circ}{v}_{n}$ denote the n neighbors of $\overset{\circ}{v}$. The new position of v is computed as

$$v = \frac{\alpha(n)\,\overset{\circ}{v} + \overset{\circ}{v}_{1} + \cdots + \overset{\circ}{v}_{n}}{\alpha(n) + n}$$

where

$$\alpha(n) = \frac{n\,(1 - \beta(n))}{\beta(n)} \quad \text{and} \quad \beta(n) = \frac{5}{4} - \frac{(3 + 2\cos(2\pi/n))^{2}}{32}$$

The function $\alpha(n)$ has been carefully chosen so that the limit surface has a continuous tangent plane [60, 73]. ■

The weights used in the averaging step are collectively referred to as a *mask*, as they are for curves. For surfaces, masks are most conveniently expressed by drawing a picture. For instance, the picture for Loop's mask is shown in Figure 10.6(a). Notice that only the numerators of the weights are shown; the denominator is always taken to be the sum of the values in the picture. This convention ensures that the weights always sum to one and therefore that the surface generated by the scheme is independent of the coordinate system used to do the calculations. The fancy term for this type of independence is *affine-invariance* [35].

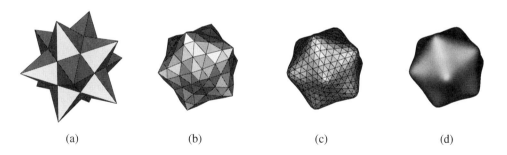

FIGURE 10.5 Loop's subdivision scheme: (a) the control mesh M^0; (b) the mesh M^1; (c) the mesh M^2; (d) the limit surface σ.

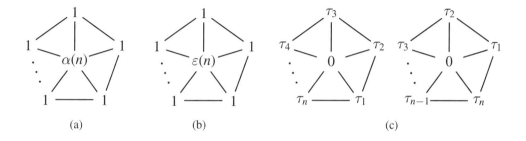

FIGURE 10.6 Masks associated with Loop's scheme: (a) the averaging mask (applied after splitting), where $\alpha(n) = n(1 - \beta(n))/\beta(n)$; (b) the evaluation mask, where $\varepsilon(n) = 3n/(4\beta(n))$; (c) the two tangent masks, where $\tau_i = \cos(2\pi i/n)$.

Evaluation masks, which can be used to send a vertex to its limit position, can also be determined for surfaces, using the same procedure as used for curves. That is, the evaluation mask for a scheme is the dominant left eigenvector of the local subdivision matrix. The evaluation mask for Loop's scheme is shown in Figure 10.6(b). A vertex v of M^j with neighbors $v_1, \ldots, v_n$ can therefore be sent to its limit position v^∞ as follows:

$$v^\infty = \frac{\varepsilon(n)\, v + v_1 + \cdots + v_n}{\varepsilon(n) + n} \quad \text{where} \quad \varepsilon(n) = \frac{3n}{4\beta(n)}$$

If we disregard the dominant left eigenvector, the two eigenvectors that are associated with the next largest eigenvalues of the local subdivision matrix give *tangent masks* defining two vectors that are tangent to the surface at the limit point. In Loop's scheme, for example, vectors in the tangent plane at v^∞ can be computed using the two masks shown in Figure 10.6(c). Note that evaluation and tangent masks, unlike averaging masks, are applied to vertices of a mesh M^j without splitting.

Evaluation and tangent masks are helpful in creating accurate shaded images of subdivision surfaces. Images such as the one shown in Figure 10.5(d) can be created as follows:

1. Subdivide the control mesh two or three times.

2. Send each vertex to its limit.

3. Compute a pair of vectors tangent to the limit surface using the tangent masks.

4. Form the normal to the surface by taking the cross-product of the tangents.

5. Shade the limit points with a Phong lighting model [38], using the computed normals.

As we mentioned in Chapter 6, subdivision schemes for curves come in two varieties: interpolating schemes and approximating schemes. The same is true for surface schemes. Polyhedral subdivision is obviously an interpolating scheme, since it produces surfaces that pass through the original control vertices. Loop's scheme, on the other hand, is an approximating scheme. Some applications require a subdivision scheme that is both interpolating *and* smooth. For that reason, Dyn et al. [33] developed the *butterfly scheme*. Like all interpolating schemes, each step of butterfly subdivision leaves the existing vertices unmoved and uses local averaging to compute new positions only for the edge midpoints introduced by splitting. The averaging mask employed by the butterfly scheme is shown in Figure 10.7. In Section 10.4 we will discuss in detail how biorthogonal wavelets can be constructed from interpolating triangular subdivision schemes such as polyhedral subdivision and the butterfly scheme.

Before leaving the basic topic of subdivision, we note that subdivision schemes for surfaces are classified as being either *uniform* or *nonuniform*, and *stationary* or *nonstationary*, just as they are for curves. A uniform scheme uses the same averaging mask for every vertex, and a stationary scheme uses the same averaging mask for every level of subdivision. A very general statement that holds for *all* subdivision schemes (triangular, quadrilateral, face, and vertex schemes alike) is that the vertex positions of M^j are linear combinations of the vertex positions of M^{j-1}. Thus, if c^{j-1} is a matrix whose i-th row consists of the x-, y-, and z-coordinates of vertex i of M^{j-1}, there exists a rectangular matrix of constants P^j such that

$$c^j = P^j c^{j-1} \tag{10.1}$$

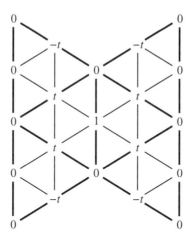

FIGURE 10.7 The averaging mask used to compute an edge midpoint in the butterfly scheme. Thick lines indicate edges of the original mesh M^{j-1}, and thin lines indicate edges introduced into $\mathring{M}^{j}$ during splitting. The weights that are zero are included only to provide context. The parameter t can be used to adjust the subdivision scheme from polyhedral ($t = 0$) to smooth ($t = 1/8$).

Just as it does for curves, the matrix P^{j} characterizes the subdivision method, no matter what type of subdivision is used.

10.2.1 Nested spaces and refinable scaling functions

We would like to show that surface subdivision can be used to define a collection of refinable scaling functions and hence a sequence of nested linear spaces. To do this, we need to decide what the domain of the scaling functions should be. In Chapter 6, we used the real line as the domain for scaling functions associated with subdivision curves; in other words, parametric curves were constructed as functions from the real line (or a bounded portion of it) into $\mathbb{R}^2$ or $\mathbb{R}^3$. A surface must be parameterized over a domain whose topological type matches that of the surface. If the surface happens to be a topological disc, then we can use a portion of $\mathbb{R}^2$ as the domain. However, $\mathbb{R}^2$ cannot be used for more general surfaces.

The approach we'll take here is to treat the initial control mesh M^0 as the domain. The initial control mesh is a reasonable candidate for the domain, as it has the same topological type as the limit surface.

In order to deduce the form of the scaling functions, we'll need to parameterize the limit surface over the domain. In general terms, a surface parameterization is nothing more than a

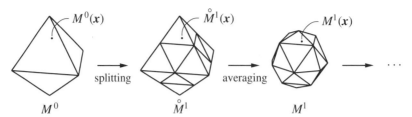

FIGURE 10.8 Tracking the progress of a point x on the base mesh M^0 in order to parameterize the limit surface $\sigma(x)$.

correspondence between points in a two-dimensional domain and points on a surface. The idea behind establishing a parameterization for a subdivision surface is to choose any point x on M^0 and track its progress through the subdivision process, as shown in Figure 10.8. The point x being tracked will converge to a point on the limit surface σ, so we may as well call this point $\sigma(x)$. Equivalently, we can establish the parameterized surface $\sigma(x)$ by using each mesh to define a piecewise-linear function $M^j(x)$, where x has M^0 as its domain. We then take $\sigma(x)$ to be the function to which the sequence of functions $M^0(x), M^1(x), \ldots$ converges.

A formal treatment of the parameterization of subdivision surfaces is given by Lounsbery et al. [75]. For our purposes, it suffices to observe that each step of the subdivision process is linear in the vertices c^j. By analogy with equation (6.7) for curves, this linearity holds even in the limit, meaning that each point $\sigma(x)$ on the limit surface can be written as a linear combination of the initial control vertices in c^0:

$$\sigma(x) = \sum_i c_i^0 \, \phi_i^0(x) \quad \text{for} \quad x \in M^0$$

The same line of reasoning used to establish the refinability of scaling functions for curves can be used to show that scaling functions for surfaces are refinable and that the subdivision matrix P^j also serves as the refinement matrix. This fact can be expressed as an equation:

$$\Phi^{j-1}(x) = \Phi^j(x) \, P^j \quad \text{for} \quad x \in M^0 \tag{10.2}$$

It follows that a single scaling function $\phi_k^j(x)$ can be computed by setting $c_i^j = \delta_{ik}$ and applying the subdivision scheme to obtain a scalar-valued function defined on the control mesh M^0. Note that even though the domain is a polyhedron, the scaling functions themselves may in fact be smooth almost everywhere (depending on the subdivision scheme).

A chain of nested linear spaces $V^j(M^0)$ associated with a mesh M^0 can now be defined as:

$$V^j(M^0) := \text{span} \left\{ \phi_1^j(\mathbf{x}), \phi_2^j(\mathbf{x}), \ldots \right\}$$

That is, $V^j(M^0)$ is the space of all functions expressible as linear combinations of the scaling functions contained in $\mathbf{\Phi}^j(\mathbf{x})$. The refinement relation given in equation (10.2) implies that these spaces are nested as required by multiresolution analysis.

10.3 Selecting an inner product

We now have almost all of the ingredients necessary to construct a biorthogonal wavelet basis for surfaces. Our goal in Section 10.4 will be to define wavelets that are nearly orthogonal to our scaling functions; therefore, we need to choose an inner product that can be used to characterize orthogonality. The inner product must be defined for functions whose domain is M^0. If we let $\Delta(M^0)$ denote the set of triangular faces of M^0 and τ denote a triangle in this set, we can define a convenient inner product as follows:

$$\langle f \mid g \rangle := \sum_{\tau \in \Delta(M^0)} \frac{1}{\text{area}\,(\tau)} \int_{\mathbf{x} \in \tau} f(\mathbf{x})\, g(\mathbf{x})\, d\mathbf{x} \tag{10.3}$$

Here $d\mathbf{x}$ corresponds to the usual differential area in $\mathbb{R}^3$.

Our definition of inner product may look strange at first, since it is independent of the geometric positions of the vertices of M^0. This choice, however, has an important practical benefit. Because inner products do not depend on the geometry of the mesh M^0, a significant amount of precomputation of inner products and wavelets can be performed—allowing the wavelet algorithms described in the next chapter to be implemented much more efficiently.

There is a potential drawback to the inner product defined above: in the process of constructing an approximation to a function, each triangle is weighted equally, independent of its true geometric size. The consequences of weighting triangles equally depend on the particular application. However, we have found no problems for the compression examples described in Section 11.2. An alternative is to define the inner product so as to weight the integral by the areas of triangles in M^0. Whether such a definition has enough important practical benefit to offset the increased computation may be an interesting topic for future research.

10.4 A biorthogonal surface wavelet construction

In this section we use the notion of lifting presented in Section 7.4 to construct a biorthogonal wavelet basis with the following properties:

- The wavelets can be constructed for surfaces generated using any triangular face scheme.

- For interpolating schemes (such as polyhedral subdivision and the butterfly scheme), analysis and synthesis are both linear-time procedures.

- The wavelets are "nearly orthogonal" to the scaling functions, in the sense that the inner product of a wavelet and a scaling function is close to zero. A practical implication is that low-resolution surface approximations are close to least-squares best.

Our construction is analogous to the construction of the single-knot wavelet in Section 7.4. We begin with lazy wavelets and then use lifting to improve them. For any triangular subdivision scheme, the lazy wavelets $\mathbf{\Psi}_{\mathrm{lazy}}^{j-1}(x)$ consist of the scaling functions in $V^j(M^0)$ associated with the midpoints of the edges of M^{j-1}. Although lazy wavelets have the advantage of being very simple, they suffer two problems:

1. While synthesis is fast, analysis is fast only for interpolating schemes, as shown later in this section. (Incidentally, the same would be true for curves, except that LU decomposition of the synthesis matrices can be used in place of the dense analysis matrices, as described in Section 7.3.2.)

2. The lazy wavelets are far from orthogonal to $V^{j-1}(M^0)$. Just as for curves, we obtain coarse surface approximations using lazy wavelets by subsampling the high-resolution coefficients. Therefore, the coarse versions of a full-resolution surface are far from least-squares best.

To address the first problem it is convenient to order the scaling functions in $\mathbf{\Phi}^j(x)$ so that the functions associated with the vertices of M^{j-1} precede the ones associated with the edge midpoints. Such an ordering is shown in Figure 10.9, where solid dots indicate vertices of M^{j-1} and open dots indicate midpoints introduced through subdivision. If we let $\mathbf{\Phi}_{\mathrm{v}}^j(x)$ and $\mathbf{\Phi}_{\mathrm{m}}^j(x)$ be the scaling functions associated with vertices and midpoints of M^{j-1}, respectively, then the ordering can be expressed in block matrix form as

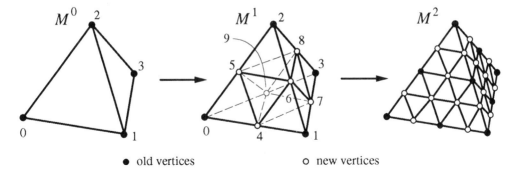

FIGURE 10.9 Polyhedral subdivision of a tetrahedron.

$$\mathbf{\Phi}^j(x) = [\ \mathbf{\Phi}_{\mathrm{v}}^j(x)\ |\ \mathbf{\Phi}_{\mathrm{m}}^j(x)\ ]$$

Example: As a running example throughout the remainder of this section, we'll use the simplest subdivision scheme and the simplest control mesh: polyhedral subdivision applied to a tetrahedron, as shown in Figure 10.9. For this example, $\mathbf{\Phi}_{\mathrm{v}}^1(x) = [\phi_0^1(x)\ \cdots\ \phi_3^1(x)]$ and $\mathbf{\Phi}_{\mathrm{m}}^1(x) = [\phi_4^1(x)\ \cdots\ \phi_9^1(x)]$. ∎

With this notation and ordering convention, the refinement relation for the lazy wavelets can be written as

$$[\ \mathbf{\Phi}^{j-1}(x)\ |\ \mathbf{\Psi}_{\mathrm{lazy}}^{j-1}(x)\ ] = [\ \mathbf{\Phi}_{\mathrm{v}}^j(x)\ |\ \mathbf{\Phi}_{\mathrm{m}}^j(x)\ ]\,[\ P_{\mathrm{lazy}}^j\ |\ Q_{\mathrm{lazy}}^j\ ]$$

where the lazy-wavelet synthesis matrix is

$$[\ P_{\mathrm{lazy}}^j\ |\ Q_{\mathrm{lazy}}^j\ ] = \begin{bmatrix} P_{\mathrm{v}}^j & \mathbf{0} \\ P_{\mathrm{m}}^j & I \end{bmatrix}$$

Example: Using the refinement relation of equation (10.2), it is easy to construct the synthesis matrices P_{lazy}^1 and Q_{lazy}^1 for the tetrahedral example shown in Figure 10.9. These matrices are the following (using the labeling indicated in the figure and dots for zeros):

$$P_{lazy}^1 = \left[\frac{P_v^1}{P_m^1}\right] = \frac{1}{2}\begin{bmatrix} 2 & \cdot & \cdot & \cdot \\ \cdot & 2 & \cdot & \cdot \\ \cdot & \cdot & 2 & \cdot \\ \cdot & \cdot & \cdot & 2 \\ \hline 1 & 1 & \cdot & \cdot \\ 1 & \cdot & 1 & \cdot \\ \cdot & 1 & 1 & \cdot \\ \cdot & 1 & \cdot & 1 \\ \cdot & \cdot & 1 & 1 \\ 1 & \cdot & \cdot & 1 \end{bmatrix} \quad Q_{lazy}^1 = \begin{bmatrix} \cdot & \cdot & \cdot & \cdot & \cdot & \cdot \\ \cdot & \cdot & \cdot & \cdot & \cdot & \cdot \\ \cdot & \cdot & \cdot & \cdot & \cdot & \cdot \\ \cdot & \cdot & \cdot & \cdot & \cdot & \cdot \\ 1 & \cdot & \cdot & \cdot & \cdot & \cdot \\ \cdot & 1 & \cdot & \cdot & \cdot & \cdot \\ \cdot & \cdot & 1 & \cdot & \cdot & \cdot \\ \cdot & \cdot & \cdot & 1 & \cdot & \cdot \\ \cdot & \cdot & \cdot & \cdot & 1 & \cdot \\ \cdot & \cdot & \cdot & \cdot & \cdot & 1 \end{bmatrix} \qquad \blacksquare$$

All blocks of the synthesis matrices are sparse, meaning that synthesis is fast. Unfortunately, the corresponding analysis matrices aren't necessarily sparse. In block form, the analysis matrices can be written thus:

$$\left[\frac{A_{lazy}^j}{B_{lazy}^j}\right] = \begin{bmatrix} P_v^j & 0 \\ P_m^j & I \end{bmatrix}^{-1} = \begin{bmatrix} (P_v^j)^{-1} & 0 \\ -P_m^j(P_v^j)^{-1} & I \end{bmatrix}$$

In Section 7.3.2, we used LU decomposition to avoid forming dense analysis matrices, but that trick won't work for surfaces because the synthesis matrices do not have the necessary banded structure. Lazy wavelet analysis will therefore be fast only when P_v^j has a sparse inverse. While this condition does not hold in general for approximating subdivision methods like Loop's scheme, it does hold for any interpolating scheme. Recall from Section 10.2 that interpolating schemes never move a vertex once it is introduced, meaning that $P_v^j = I$. The synthesis and analysis matrices for interpolating lazy wavelets are therefore always sparse, and are given by

$$[\, P_{lazy}^j \mid Q_{lazy}^j \,] = \begin{bmatrix} I & 0 \\ P_m^j & I \end{bmatrix} \quad \text{and} \quad \left[\frac{A_{lazy}^j}{B_{lazy}^j}\right] = \begin{bmatrix} I & 0 \\ -P_m^j & I \end{bmatrix}$$

Example: The analysis matrices A_{lazy}^1 and B_{lazy}^1 for the polyhedral subdivision case shown in Figure 10.9 are

$$A_{\text{lazy}}^1 = \begin{bmatrix} 1 & \cdot & \cdot & \cdot & \cdot & \cdot & \cdot & \cdot \\ \cdot & 1 & \cdot & \cdot & \cdot & \cdot & \cdot & \cdot \\ \cdot & \cdot & 1 & \cdot & \cdot & \cdot & \cdot & \cdot \\ \cdot & \cdot & \cdot & 1 & \cdot & \cdot & \cdot & \cdot \end{bmatrix}$$

$$B_{\text{lazy}}^1 = \frac{1}{2} \begin{bmatrix} -1 & -1 & \cdot & \cdot & 2 & \cdot & \cdot & \cdot & \cdot & \cdot \\ -1 & \cdot & -1 & \cdot & \cdot & 2 & \cdot & \cdot & \cdot & \cdot \\ \cdot & -1 & -1 & \cdot & \cdot & \cdot & 2 & \cdot & \cdot & \cdot \\ \cdot & -1 & \cdot & -1 & \cdot & \cdot & \cdot & 2 & \cdot & \cdot \\ \cdot & \cdot & -1 & -1 & \cdot & \cdot & \cdot & \cdot & 2 & \cdot \\ -1 & \cdot & \cdot & -1 & \cdot & \cdot & \cdot & \cdot & \cdot & 2 \end{bmatrix} \qquad \blacksquare$$

To construct wavelets that are more orthogonal to scaling functions than the lazy wavelets, we can proceed much as we did for the construction of the single-knot wavelets in Section 7.4. Once again, the idea is to modify a lazy wavelet by subtracting a linear combination of nearby coarse scaling functions. The i-th lazy wavelet at level $j - 1$ is just the i-th scaling function in $\boldsymbol{\Phi}_{\text{m}}^j$, which we will denote $\phi_{\text{m},i}^j$. We define an improved wavelet as follows:

$$\psi_i^{j-1}(\boldsymbol{x}) = \phi_{\text{m},i}^j(\boldsymbol{x}) - \sum_k s_{k,i}^j \, \phi_k^{j-1}(\boldsymbol{x})$$

where k is restricted to a few values corresponding to vertices in M^{j-1} in the neighborhood of $\phi_{\text{m},i}^j$. The values s_{ik}^j are determined for the i-th wavelet by solving, in a least-squares sense, the system of equations

$$\langle \, \psi_i^{j-1} \mid \phi_{i'}^{j-1} \rangle = 0 \tag{10.4}$$

for all i' such that the support of $\psi_i^{j-1}(\boldsymbol{x})$ overlaps with $\phi_{i'}^{j-1}(\boldsymbol{x})$.

We turn now to the question of which values $s_{k,i}^j$ should be allowed to be nonzero for each i. Recall that $\psi_i^{j-1}(\boldsymbol{x})$ is associated with the midpoint of an edge of M^{j-1}; call the two endpoints of that edge the "parent" vertices. The smallest symmetric choice for the column vector S_i^j therefore consists of just two nonzero entries corresponding to the two parents. A symmetric way to increase the supports is to add entries to S_i^j corresponding to the k-discs of the parent vertices. (The k-disc of a vertex is the set of all vertices reachable by following k or fewer edges of the triangulation.) Some examples of k-discs of two edge endpoints are shown in Figure 10.10. As the number of nonzero entries of S_i^j is allowed to grow, the supports of the wavelets $\boldsymbol{\Psi}^{j-1}$ grow, and the wavelets become increasingly orthogonal to $V^{j-1}(M^0)$.

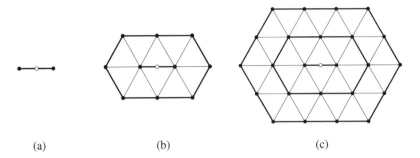

FIGURE 10.10   Vertices within the k-discs of two edge endpoints in a regular triangulation: (a) the 0-disc of the endpoints; (b) the 1-disc of the endpoints; (c) the 2-disc of the endpoints.

Once we choose a k-disc of vertices and the corresponding scaling functions that will be used to modify the lazy wavelet, the last step in determining a wavelet is to solve equation (10.4) in the least-squares sense for the unknown values of S_i^j. We'll refer to the wavelets constructed in this way as *k-disc wavelets* and use the subscript "kd" to distinguish them from lazy wavelets. Some typical examples of polyhedral wavelets are shown in Figure 10.11.

The k-disc wavelet synthesis and analysis matrices $P_{kd}^j, Q_{kd}^j, A_{kd}^j, B_{kd}^j$ are obtained from the lazy wavelet matrices by combining all the column vectors S_i^j into a matrix S^j and using this matrix to perform lifting in the same manner as described by equation (7.26):

$$[\, P_{kd}^j \mid Q_{kd}^j \,] = [\, P_{lazy}^j \mid Q_{lazy}^j - P_{lazy}^j\, S^j \,]$$

$$\left[\begin{array}{c} A_{kd}^j \\ \hline B_{kd}^j \end{array}\right] = \left[\begin{array}{c} A_{lazy}^j + S^j\, B_{lazy}^j \\ \hline B_{lazy}^j \end{array}\right]$$

Example: The tetrahedron shown in Figure 10.9 has so few vertices that the wavelets Ψ^1 constructed using 1-discs have global support (wavelets at resolution $j > 1$ will, however, be locally supported). The matrix S^1 is therefore dense:

$$S^1 = \frac{1}{8}\begin{bmatrix} 3 & 3 & -1 & -1 & -1 & 3 \\ 3 & -1 & 3 & 3 & -1 & -1 \\ -1 & 3 & 3 & -1 & 3 & -1 \\ -1 & -1 & -1 & 3 & 3 & 3 \end{bmatrix}$$

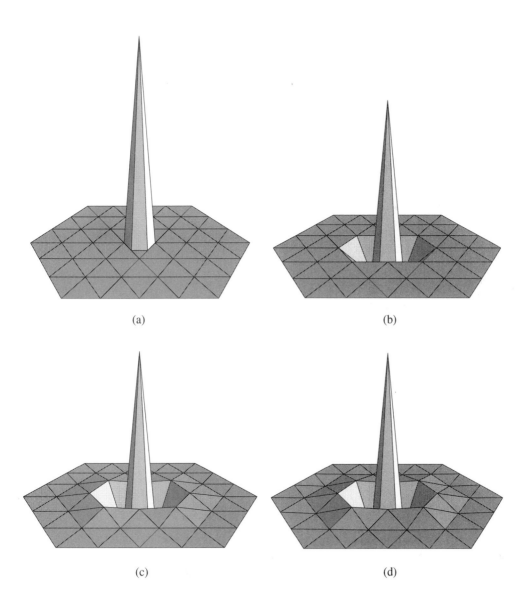

(a)

(b)

(c)

(d)

FIGURE 10.11 Polyhedral surface wavelets: (a) a lazy wavelet; (b) a 0-disc wavelet; (c) a 1-disc wavelet; (d) a 2-disc wavelet.

The corresponding analysis and synthesis matrices are

$$
P^1_{\text{kd}} = \frac{1}{2}\begin{bmatrix} 2 & \cdot & \cdot & \cdot \\ \cdot & 2 & \cdot & \cdot \\ \cdot & \cdot & 2 & \cdot \\ \cdot & \cdot & \cdot & 2 \\ \hline 1 & 1 & \cdot & \cdot \\ 1 & \cdot & 1 & \cdot \\ \cdot & 1 & 1 & \cdot \\ \cdot & 1 & \cdot & 1 \\ \cdot & \cdot & 1 & 1 \\ 1 & \cdot & \cdot & 1 \end{bmatrix}
\qquad
Q^1_{\text{kd}} = \frac{1}{8}\begin{bmatrix} -3 & -3 & 1 & 1 & 1 & -3 \\ -3 & 1 & -3 & -3 & 1 & 1 \\ 1 & -3 & -3 & 1 & -3 & 1 \\ 1 & 1 & 1 & -3 & -3 & -3 \\ 5 & -1 & -1 & -1 & 1 & -1 \\ -1 & 5 & -1 & 1 & -1 & -1 \\ -1 & -1 & 5 & -1 & -1 & 1 \\ -1 & 1 & -1 & 5 & -1 & -1 \\ 1 & -1 & -1 & -1 & 5 & -1 \\ -1 & -1 & 1 & -1 & -1 & 5 \end{bmatrix}
$$

$$
A^1_{\text{kd}} = \frac{1}{16}\begin{bmatrix} 7 & -1 & -1 & -1 & 6 & 6 & -2 & -2 & -2 & 6 \\ -1 & 7 & -1 & -1 & 6 & -2 & 6 & 6 & -2 & -2 \\ -1 & -1 & 7 & -1 & -2 & 6 & 6 & -2 & 6 & -2 \\ -1 & -1 & -1 & 7 & -2 & -2 & -2 & 6 & 6 & 6 \end{bmatrix}
$$

$$
B^1_{\text{kd}} = \frac{1}{2}\begin{bmatrix} -1 & -1 & \cdot & \cdot & 2 & \cdot & \cdot & \cdot & \cdot & \cdot \\ -1 & \cdot & -1 & \cdot & \cdot & 2 & \cdot & \cdot & \cdot & \cdot \\ \cdot & -1 & -1 & \cdot & \cdot & \cdot & 2 & \cdot & \cdot & \cdot \\ \cdot & -1 & \cdot & -1 & \cdot & \cdot & \cdot & 2 & \cdot & \cdot \\ \cdot & \cdot & -1 & -1 & \cdot & \cdot & \cdot & \cdot & 2 & \cdot \\ -1 & \cdot & \cdot & -1 & \cdot & \cdot & \cdot & \cdot & \cdot & 2 \end{bmatrix}
$$

■

10.5 Multiresolution representations of surfaces

The k-disc wavelets presented in the previous section define a multiresolution basis $[\Phi^0(x) \; \Psi^0_{\text{kd}}(x) \; \Psi^1_{\text{kd}}(x) \; \cdots \; \Psi^{J-1}_{\text{kd}}(x)]$ for the approximation space $V^J(M^0)$. A surface $\sigma(x)$ parameterized on a simple mesh M^0 can therefore be expanded, at least approximately, in this basis. The surface can be written exactly if it happens to lie in $V^J(M^0)$; otherwise, the error can typically be made arbitrarily small by increasing J. (We say "typically" here because one could imagine pathological cases in which either the surface σ or the subdivision scheme is sufficiently ill-behaved to keep the approximation error from decreasing.)

A multiresolution representation of a surface $\sigma(x)$ parameterized on M^0 thus consists of a simple coarse surface together with wavelet coefficients at various scales. The coarse surface $\sigma^0(x)$ is the projection of $\sigma(x)$ into the lowest-resolution space $V^0(x)$. If polyhedral subdivision is used, then $\sigma^0(x)$ will be a coarse polyhedron that is structurally equivalent to M^0. If a smooth subdivision scheme is used, then $\sigma^0(x)$ will be a smooth approximation to the surface $\sigma(x)$. The example of Spock's head in Figure 10.1 illustrates the decomposition of a polyhedral subdivision surface that is parameterized on an octahedron; thus, the coarsest approximation is an octahedron whose vertices are positioned so that the least-squared error between σ^0 and σ is small. If we had used the butterfly scheme instead of polyhedral subdivision in Figure 10.1, each of the approximations to Spock's head would be a smooth subdivision surface.

One way to expand a surface $\sigma(x)$ onto the k-disc wavelet basis is to use the dual basis $[\widetilde{\boldsymbol{\Phi}}^0(x) \quad \widetilde{\boldsymbol{\Psi}}^0_{kd}(x) \quad \widetilde{\boldsymbol{\Psi}}^1_{kd}(x) \quad \cdots \quad \widetilde{\boldsymbol{\Psi}}^{J-1}_{kd}(x)]$, as described in Section 7.4. For instance, a particular wavelet coefficient d_i^j can be computed using an inner product with the normalized dual wavelet $\widetilde{\psi}_i^j$ or unnormalized dual wavelet $\overline{\psi}_i^j$ as follows:

$$d_i^j = \langle \, \widetilde{\psi}_i^j(x) \mid \sigma(x) \, \rangle$$
$$= 4^j \langle \, \overline{\psi}_i^j(x) \mid \sigma(x) \, \rangle$$

In the next chapter, we describe an alternative method of computing the k-disc basis coefficients for a given surface, based on the use of a filter-bank algorithm. The next chapter also presents various applications of the k-disc wavelet representations, including surface compression, level-of-detail control, progressive transmission, and multiresolution surface editing.

SURFACE APPLICATIONS

1. Conversion to multiresolution form — 2. Surface compression —
3. Continuous level-of-detail control — 4. Progressive transmission —
5. Multiresolution editing — 6. Future directions for surface wavelets

In this chapter, we explore various uses of the k-disc surface wavelets developed in Chapter 10. These applications include compression of surface models, continuous level-of-detail control for high-performance rendering, progressive transmission of complex surface models, and multiresolution editing of surfaces.

Most of the surfaces encountered in practice are currently represented in a more traditional form—for example, as complicated polyhedra, or as B-spline surface patches. Therefore, we often need to convert these other surface representations to multiresolution form. We begin the chapter with a discussion of this conversion process.

11.1 Conversion to multiresolution form

In Section 10.5 we showed how a surface $\sigma(x)$ parameterized on a simple mesh M^0 could be expanded, at least approximately, onto a basis of piecewise-linear k-disc wavelets using biorthogonal duals. The principal difficulty for practical application of this process is that in most situations, neither the simple mesh M^0 nor the parameterization of the surface is known. For

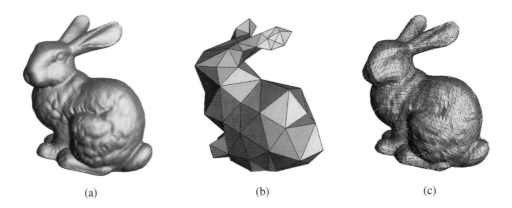

FIGURE 11.1   (a) A complex mesh consisting of approximately 70,000 triangles created using an optical scanner and the zippering technique of Turk and Levoy [125]; (b) the base mesh M^0 constructed by the algorithm of Eck et al. [34]; (c) the projection of the model into $V^5(M^0)$.

instance, the bunny shown in Figure 11.1(a) is initially defined only as a collection of nearly 70,000 triangles stitched together into a complicated mesh.

The first step of converting an arbitrary surface into multiresolution form is therefore to determine a simple mesh M^0 (called a *base mesh*) that is topologically equivalent to the given surface and a parametric function $\sigma(x)$ that maps points $x \in M^0$ into three-dimensional space. An algorithm for solving this problem has recently been developed by Eck et al. [34] for the case in which the initial surface is a polyhedron. Their algorithm produces the base mesh M^0 for the bunny shown in Figure 11.1(b). Parameterizing complex surfaces that are not polyhedral remains an open problem.

Once a parameterization is constructed for a given polyhedral surface, Eck et al. go on to convert the surface to multiresolution form. This conversion is accomplished by projecting the initial surface into an approximation space $V^J(M^0)$, where J is determined so that the maximum error is below a user-specified tolerance. The projection step amounts to recursively subdividing the base mesh J times. The parameterization for the surface is then used to map the newly generated vertices into three-dimensional space. The mapped vertices describe a polyhedron in $V^J(M^0)$, as illustrated in Figure 11.1(c) for the bunny at level $J = 5$. The final step of the conversion to multiresolution form is the application of filter-bank analysis using the matrices A_{kd}^j and B_{kd}^j, which yields a coarse surface approximation and a collection of wavelet coefficients.

11.2 Surface compression

In this section, k-disc wavelets are used to address two compression applications: compression of complex surfaces and compression of texture maps defined on surfaces.

11.2.1 Polyhedral compression

The first application of k-disc wavelets is to compress polyhedral models such as the one shown in Color Plate 9(a). This particular model, consisting of 32,768 triangles, was created from range data provided by Cyberware, Inc. Since the original data was gridded and the surface was known to be topologically equivalent to a sphere, conversion to multiresolution form did not require the general parameterization algorithm of Eck et al. Instead, a special-purpose procedure was used to parameterize the model on an octahedral base mesh [74]. The surface was then converted to multiresolution form using recursive subdivision followed by filter-bank analysis, as described above.

The wavelet coefficients computed by the filter-bank algorithm are coefficients of un-normalized basis functions, so the magnitude of a coefficient is not a good measure of the least-squares error that would result if that coefficient were removed. If we multiply each wavelet coefficient d_i^j at level j by 2^{-j}, we get coefficients for an L^2 normalized basis. These normalized coefficients have magnitudes that are meaningful in a compression algorithm like the one discussed for images in Chapter 3.

The surface approximations shown in Color Plate 9(c), (f), and (i) were computed by sorting the normalized coefficients of the 2-disc piecewise-linear wavelets, then removing 99%, 88%, and 70% of the smallest-magnitude coefficients, respectively. Notice that this simple strategy causes the approximation to refine more deeply in areas of high detail while leaving large triangles in areas of relatively low detail. The same technique was applied to the bunny model, with the results depicted in Color Plate 10.

Although all our examples of k-disc wavelets up to this point have used polyhedral subdivision, the multiresolution framework for surfaces is general enough to encompass smooth subdivision schemes as well. Recall from Section 10.4 that butterfly subdivision, being an interpolating scheme, allows for linear-time wavelet analysis. We can therefore use k-disc wavelets to approximate Spock's head using smooth surfaces with varying amounts of detail, as illustrated in Color Plate 11. The fully detailed surface is shown in Color Plate 11(a), and a compressed representation that requires only 16% of the coefficients of the original surface is shown in Color Plate 11(b). The compression was performed using 2-disc wavelets constructed from the butterfly scheme.

11.2.2   Texture map compression

Compression can also be applied to the representation of a texture map defined on a surface. If the surface is parameterized on the unit square, a texture map is no different from an ordinary image and hence can be compressed using traditional wavelet techniques such as Haar wavelets or spline wavelets. However, if the surface is topologically more complicated, traditional wavelets no longer suffice, but k-disc wavelets can be used. The idea is to treat each color component—red, green, and blue—as a scalar function defined on the base mesh M^0. Each of the color functions can be converted to multiresolution form using filter-bank analysis, and the normalized wavelet coefficients can then be sorted by magnitude and truncated just as in image compression.

In the example shown in Color Plate 12, elevation and bathymetry data obtained from the U.S. National Geophysical Data Center were used to create a piecewise-linear coloring of the globe. The resulting color function contains 2,097,152 triangles and 1,048,578 vertices. The full-resolution coloring was too large to be rendered on a graphics workstation with 128 megabytes of memory and is therefore not shown in its entirety in Color Plate 12. You can, however, get an appreciation for the density of the data from Color Plate 12(h), where even at close range the mesh lines are so closely spaced that the image is almost completely black.

The approximations shown in Color Plate 12(a) through (f) were produced by leaving out 2-disc piecewise-linear wavelet coefficients with magnitudes smaller than a certain threshold. Color Plate 12(a) shows a distant view of the earth in which 99.9% of the wavelet coefficients have been eliminated, with the corresponding mesh shown in (b). Similarly, Color Plate 12(c) and (d) show the results of eliminating 98% of the coefficients for a medium-range view. At close range, the 90%-compressed model in (e) is nearly indistinguishable from the full resolution model in (g). A comparison of the compressed mesh in (f) and the original in (h) reveals the striking degree of compression achieved in this case.

11.3   Continuous level-of-detail control

When viewing a complex object, it is unnecessary and inefficient to draw a highly detailed representation if the viewer is far away from the object. Instead, we would like to use some form of *level-of-detail control*—allowing information about the view to determine the complexity of the model that is rendered. Currently, perhaps the most common approach to creating LOD models is to have the user craft them by hand. In contrast, the surface compression technique described in Section 11.2 provides a mechanism for *automatically* producing LOD models.

The images in Color Plate 9 illustrate the use of wavelet approximations for automatic level-of-detail control in rendering. The left-hand column of images shows the full-resolution mesh as viewed from various distances. When viewing the original polyhedron from the more distant vantage points, there is no need to render all 32,000 triangles. The approximations shown in the second column may be used instead, without significantly degrading the quality of the resulting images.

Switching suddenly between models with different levels of detail in an animation can produce objectionable "popping." This problem is easily solved by using continuous levels of smoothing in the same manner as discussed for curves in Section 8.2. In effect, as the viewer approaches an object, each wavelet coefficient is smoothly varied from zero to its correct value. Likewise, as the viewer recedes, each wavelet coefficient is smoothly reduced to zero. More generally, each wavelet coefficient can be made a continuous function of the viewing distance.

11.4 Progressive transmission

Text, images, and video are commonplace on the World Wide Web, and complex geometric models are becoming very common as well. The ever-growing production and distribution of these geometric objects motivates the need for efficient transmission of models across relatively low-bandwidth networks.

The most straightforward way to transmit a mesh that defines a subdivision surface is by sending each of the triangles of the highest-resolution representation over the network. However, transmitting complex meshes in this way forces the user to wait until the entire model is received before anything can be displayed. A more attractive alternative is to use a wavelet representation for *progressive transmission*, as illustrated in Figure 11.2. First, the base mesh is transmitted; since this mesh contains very few triangles, it is received and displayed quickly. Next, the normalized wavelet coefficients are transmitted in order of decreasing magnitude. As these coefficients are received, the renderer can update and redisplay the model.

11.5 Multiresolution editing

The multiresolution representation of curves discussed in Chapter 8 supported a variety of editing operations. In much the same way, surface wavelets can be used to edit shapes at a variety of resolutions. We have already mentioned one example of multiresolution surface editing at the beginning of Chapter 10, where we discussed the editing of tensor-product surfaces, as shown in Color Plate 8.

We can also edit surfaces of arbitrary topological type in much the same way. A simple example is given in Color Plate 11(c) and (d). Color Plate 11(c) shows the effect of changing a single scaling function coefficient of the level-0 base octahedron of Spock's head. Because

(a)   (b)   (c)

FIGURE 11.2   Progressive transmission: (a) the base mesh, consisting of 229 triangles; (b) the mesh after approximately 2,000 wavelet coefficients have been received; (c) the mesh after approximately 10,000 wavelet coefficients have been received. (Original model courtesy of Greg Turk and Marc Levoy.)

finer-level vertices in the same region are defined relative to the coarser shape, they move along with the modification. However, the geometry in areas away from the front of the bust is not affected. It is also possible to modify the shape locally by changing the value of a wavelet coefficient at a finer level. The result of modifying a single level-3 wavelet coefficient is shown in Color Plate 11(d).

The surface editing examples in Color Plate 11 were created by simply modifying a single value in the wavelet representation. Of course, in order to make more general modifications, it would be valuable to have a powerful surface-editing tool with capabilities similar to those described for multiresolution curve editing in Section 8.3 and in the work by Finkelstein and Salesin [37].

11.6   Future directions for surface wavelets

The application of multiresolution analysis to surfaces of arbitrary topology is a relatively recent development in the graphics community. The k-disc wavelets that we have developed in the previous chapter and applied in this one are certainly not the last word in multiresolution surface representations. The representation described could be enhanced in many ways:

- In the methods we've described, the wavelet decomposition of a surface always retains the topological type of the input surface. However, when the input is a relatively simple object with many small holes, it might be desirable to decompose the input into a topologically simpler surface, that is, one with lower genus or fewer boundary curves.

- The images in Color Plates 9 and 12 were generated by simply incorporating the wavelet coefficients of greatest magnitude. A view-dependent error metric could be used to produce images of better quality using even fewer triangles.

- For surfaces created from physical objects using range-scanning it is possible to represent the shape and the color as separate wavelet expansions, as described by Certain et al. [10]. Such representations may prove useful for efficiently manipulating and editing complex surfaces in an interactive viewer or editor.

IV

PHYSICAL
SIMULATION

VARIATIONAL MODELING

1. Setting up the objective function — 2. The finite-element method — 3. Using finite elements in variational modeling — 4. Variational modeling using wavelets — 5. Adaptive variational modeling

In the previous chapters, we've looked at several approaches for editing curves and surfaces in which the user manipulates the geometric primitive explicitly, either by moving control points or through "direct manipulation." And we've seen how wavelets can be used to facilitate these operations by providing a natural and efficient control over the locality of the edits.

However, in the design of curves and surfaces, it is often useful for the user to exercise control indirectly by specifying just a goal function and a few constraints and then letting the computer solve for the "best" geometric primitive that meets those constraints. Such a design process is called *variational modeling* [130], as the goal function (or *objective function*, as it is more technically called) is typically specified as the minimum of some integral, and minimizing integrals is the domain of variational calculus.

As we'll see in this chapter, it turns out that wavelets are also very useful in speeding the computations required for variational modeling. To see how wavelets can be used, we will begin by formulating the variational modeling problem more precisely and examining how finite elements are used in its solution.

12.1   Setting up the objective function

Here is an example of a variational modeling problem. Suppose we want to design a "smooth" curve that passes through some particular points in three-dimensional space. How might we do it?

The first step is to define the objective function more precisely. We'll begin by writing an expression for the curve in three-dimensional space as

$$\gamma(t) := [\gamma_1(t) \ \ \gamma_2(t) \ \ \gamma_3(t)]^{\mathrm{T}} \quad \text{for } t \in [0, 1]$$

Next we must specify the objective function. One measure of "smoothness" is the *total curvature:* the integral of the curvature over the length of the curve. A slightly simpler model—the integral of the second derivative squared, which is based on the energy of a bent wire ([99], page 248)—can also be used as a fairly good measure of curvature. The variational modeling problem then becomes

$$\text{minimize} \quad \int_0^1 |\gamma_i''(t)|^2 dt \text{ subject to the constraints, for } i = 1, 2, 3$$

Before we solve this problem, let's set up a second example—this time for surfaces. Suppose, as before, that we want to design a "smooth" surface subject to certain constraints. We can write a parametric surface as

$$\sigma(s, t) := [\sigma_1(s, t) \ \ \sigma_2(s, t) \ \ \sigma_3(s, t)]^{\mathrm{T}} \quad \text{for } s, t \in [0, 1]$$

If we approximate the total curvature of this surface by a thin-plate energy model ([99], page 318), we can once again treat each coordinate function independently. The variational modeling problem for surfaces can then be written as

$$\text{minimize} \quad \int_0^1 \int_0^1 \left(\left| \frac{\partial^2 \sigma_i}{\partial s^2} \right|^2 + 2 \left| \frac{\partial^2 \sigma_i}{\partial s \partial t} \right|^2 + \left| \frac{\partial^2 \sigma_i}{\partial t^2} \right|^2 \right) ds \, dt$$

subject to constraints, for $i = 1, 2, 3$. The surface can be constrained to pass through a point by specifying σ at a particular s and t. The normal of the surface can also be constrained by specifying $\partial \sigma / \partial s$ and $\partial \sigma / \partial t$ for some s and t.

As mentioned earlier, finding the curve or surface with minimum total curvature that satisfies a given set of constraints is a problem of variational calculus because the unknowns are functions. To make the problem computationally tractable, we need to reduce the problem to one in which the unknowns are a finite set of numbers.

12.2 The finite-element method

The *finite-element method* is a technique for finding an approximate solution to a problem by replacing the unknown function in question with an unknown linear combination of known basis functions.

The basic steps in the finite-element method are as follows:

1. Choose a set of basis functions $u(t) = [u_1(t) \cdots u_m(t)]$, called the *finite elements*.

2. Write the unknown function as a linear combination of these finite elements. For example, a function $\gamma(t)$ would be written

$$\gamma(t) = \sum_{j=1}^{m} x_j u_j(t) = u(t)\, x$$

where $x = [x_1 \cdots x_m]^{\mathrm{T}}$ is a set of unknown coefficients.

Note that the unknown function $\gamma(t)$ can only be expressed exactly in this finite-element form if it lies in the vector space spanned by the finite elements. Otherwise, the finite-element form $u(t)\, x$ can only represent the unknown function approximately.

3. Substitute the finite-element representation $u(t)\, x$ into the original problem involving $\gamma(t)$. In many cases, the original problem is easier to solve in terms of the finite set of unknowns x after this substitution.

In the next section, we'll look at how this method can be applied to variational modeling. Later, in the next chapter, we'll look at another application of finite elements to solving the equations of radiative transport for global illumination.

12.3 Using finite elements in variational modeling

We will examine how finite elements can be used in variational modeling by continuing the previous example of finding a smooth curve $\gamma(t)$ that interpolates some specified points. Here is an instance of one such variational problem:

$$\text{minimize} \quad \int_0^1 |\gamma''(t)|^2 \, dt \quad \text{subject to} \quad \begin{cases} \gamma(0) = 3 \\ \gamma(\tfrac{1}{2}) = 18 \\ \gamma(1) = 12 \end{cases}$$

To express this problem using finite elements, we will first restrict the class of curves we are willing to consider so that our solution curve $\gamma(t)$ can be written as a linear combination of a finite set of basis functions.

For instance, if we restrict $\gamma(t)$ to be a quartic function, then we can express it as follows:

$$\gamma(t) = x_1 + x_2 t + x_3 t^2 + x_4 t^3 + x_5 t^4 = u(t) \, x$$

where $u(t) = [1 \ \ t \ \ t^2 \ \ t^3 \ \ t^4]$ and $x = [x_1 \ x_2 \ x_3 \ x_4 \ x_5]^{\text{T}}$.

In this case, the five monomial basis functions in $u(t)$ are called the finite elements, and we have limited the solution space for $\gamma(t)$ to the five-dimensional space that they span.

Note, however, that there are still an infinite number of quartic functions that satisfy the given constraints (as shown in Figure 12.1), and only one of these functions has minimum curvature. To find this function, we'll first plug the finite-element representation of $\gamma(t)$ into the constraint equations of the initial problem. In this case, we get

$$\begin{bmatrix} \gamma(0) \\ \gamma(\tfrac{1}{2}) \\ \gamma(1) \end{bmatrix} = \begin{bmatrix} u(0) \\ u(\tfrac{1}{2}) \\ u(1) \end{bmatrix} x = \begin{bmatrix} 1 & 0 & 0 & 0 & 0 \\ 1 & \tfrac{1}{2} & \tfrac{1}{4} & \tfrac{1}{8} & \tfrac{1}{16} \\ 1 & 1 & 1 & 1 & 1 \end{bmatrix} x = \begin{bmatrix} 3 \\ 18 \\ 12 \end{bmatrix}$$

The result is a set of linear equations in the unknown coefficients:

$$A x = b$$

where

$$A = \begin{bmatrix} 1 & 0 & 0 & 0 & 0 \\ 1 & \tfrac{1}{2} & \tfrac{1}{4} & \tfrac{1}{8} & \tfrac{1}{16} \\ 1 & 1 & 1 & 1 & 1 \end{bmatrix} \quad \text{and} \quad b = \begin{bmatrix} 3 \\ 18 \\ 12 \end{bmatrix}$$

We can also substitute the finite-element representation of $\gamma(t)$ into the objective function. Plugging it in and performing some algebraic manipulations yields

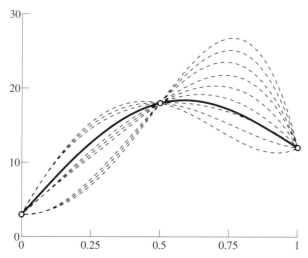

FIGURE 12.1 Quartic functions satisfying three interpolatory constraints. The solid curve shows the function with minimum curvature.

$$\int_0^1 |\gamma''(t)|^2 dt = \int_0^1 (u''(t)\,x)\,(u''(t)\,x)\,dt$$

$$= \int_0^1 (x^T u''^T)\,(u''\,x)\,dt$$

$$= \int_0^1 x^T\,(u''^T\,u'')\,x\,dt$$

$$= x^T\left(\int_0^1 u''^T\,u''\,dt\right)x$$

$$= \frac{1}{2}x^T\,H\,x$$

where the factor of $1/2$ has been introduced for later convenience, and the *Hessian matrix* H is given by

$$H = 2\int_0^1 u''(t)^{\mathrm{T}} u''(t)\, dt = \frac{4}{5}\begin{bmatrix} 0 & 0 & 0 & 0 & 0 \\ 0 & 0 & 0 & 0 & 0 \\ 0 & 0 & 10 & 15 & 20 \\ 0 & 0 & 15 & 30 & 45 \\ 0 & 0 & 20 & 45 & 72 \end{bmatrix}$$

Our original variational modeling problem is now of the form

$$\text{minimize} \quad \tfrac{1}{2} x^{\mathrm{T}} H x \quad \text{subject to} \quad A x = b$$

Problems like this one, with a quadratic objective function and linear constraints, are the subject of an area of optimization known as *quadratic programming*. To solve this problem, we'll use the *method of Lagrange multipliers* ([51], pages 43–49), which works particularly well for solving quadratic programming problems.

For an intuitive idea of how this method works, think of the function being minimized as defining a potential. In this case, the gradient of the objective function defines a force field. We're looking for the "stable point," that is, the point where the force is zero.

Following in this line of intuition, you could think of each constraint as having a force associated with it that pulls toward the "constraint surface"—the locus of points that satisfy that constraint. The difficulty with this conception is that there isn't a natural way to weight the constraint forces relative to one another or to the potential force produced by the objective function. Therefore, we'll leave the relative weights of these forces as unknowns $\lambda_1, \ldots, \lambda_m$ (these are the *Lagrange multipliers*). We now need to find weights and a stable point such that (a) the total force is zero and (b) the point is simultaneously located on all constraint surfaces. The reason this technique works so well for quadratic minimization problems is that conditions (a) and (b) together become a system of as many constraints as unknowns, and we are left with a square linear system to solve.

Let's see how this method works by completing our example problem. Since in this case we have three constraints, we'll introduce a vector of three Lagrange multipliers $\lambda = [\lambda_1\ \lambda_2\ \lambda_3]^{\mathrm{T}}$. We can now include the constraints into part of a single unconstrained minimization problem:

$$\text{minimize} \quad \tfrac{1}{2} x^{\mathrm{T}} H x + (A x - b)^{\mathrm{T}} \lambda$$

We can find the minimum by taking the derivative of this new objective function with respect to each of the unknowns x_i and λ_i and setting these derivatives equal to zero. The result is a system of linear equations:

$$\begin{cases} Hx + A^{\mathrm{T}}\lambda & = 0 \\ Ax & -b = 0 \end{cases}$$

Then we can write this system as a single matrix equation using block matrix notation:

$$\begin{bmatrix} H & A^{\mathrm{T}} \\ A & 0 \end{bmatrix} \begin{bmatrix} x \\ \lambda \end{bmatrix} = \begin{bmatrix} 0 \\ b \end{bmatrix} \tag{12.1}$$

When we plug in the values for H, A, and b and then solve for x and λ, we get

$$x = [3 \; 41 \; 8 \; -80 \; 40]^{\mathrm{T}}$$
$$\lambda = [256 \; -512 \; 256]^{\mathrm{T}}$$

In this case, we will not use the values of the Lagrange multipliers. However, the values of x tell us that the quartic curve that goes through the constraint points while minimizing total curvature is

$$\gamma(t) = 3 + 41t + 8t^2 - 80t^3 + 40t^4$$

In general, except in the most trivial cases, curves and surfaces require enough basis functions to make the matrix system in equation (12.1) quite large. However, most commonly used basis functions (such as B-splines) overlap very few of their neighbors, making H sparse. The large, sparse linear system that results is generally solved most efficiently using an iterative solution technique like the Gauss-Seidel method or the conjugate gradient method [97].

12.4 Variational modeling using wavelets

So how do wavelets fit in? Until now, we haven't specified which basis functions should be used as finite elements for solving variational problems in modeling curves and surfaces (although we did run through a simple example with monomial basis functions as finite elements).

An obvious choice to use for curves and surfaces are the B-splines, since they are already commonly used as a basis in modeling systems. However, the matrix in equation (12.1) turns out to be poorly conditioned for a B-spline basis. Intuitively, this poor conditioning is due to the fact that each B-spline basis function represents only a small portion of the solution, thereby requiring any broad changes in the curve to propagate incrementally from one basis

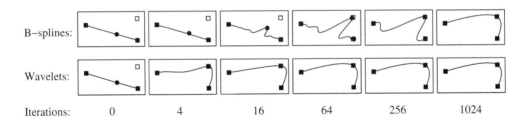

B–splines:

Wavelets:

Iterations: 0   4   16   64   256   1024

FIGURE 12.2   A sequence of iterations converging to the minimum-energy curve satisfying three constraints. The top row illustrates the slow convergence that results from using a B-spline basis; the bottom row shows the improved convergence of a wavelet basis.

function to the next. In this case, the iterative solution techniques commonly used for large sparse systems are slow to converge to the correct solution.

The slow convergence of the B-spline basis is illustrated in the upper row of Figure 12.2. This figure shows how a variational solution for the minimal-energy curve gradually converges over 1024 iterations. Clearly, use of a B-spline basis in an interactive design tool may result in a long delay as the system recomputes a new minimum-energy solution whenever the user adds or changes a constraint of the curve.

One solution to the problem of poor conditioning and slow convergence is to use wavelets constructed from B-splines (as described in Section 7.3.2) for the finite elements, rather than the B-splines themselves. This change of basis can dramatically improve the performance of the minimization process, as shown by Gortler and Cohen [49]. Intuitively, the wavelet basis allows changes in the curve to propagate much more quickly from one region to another by allowing the effects of a change to "bubble up" the hierarchy to basis functions with broader support and then descend back down the hierarchy to the narrower basis functions for the region affected.

Another way to think of this switch from the scaling functions to the wavelet basis, for those familiar with iterative solution techniques, is as a preconditioning step. Instead of solving equation (12.1) directly, we solve

$$\begin{bmatrix} W^{-T}HW^{-1} & W^{-T}A^T \\ AW^{-1} & 0 \end{bmatrix} \begin{bmatrix} \hat{x} \\ \lambda \end{bmatrix} = \begin{bmatrix} 0 \\ b \end{bmatrix}$$

where W represents the wavelet transform and $\hat{x} = Wx$ is the set of wavelet coefficients for the solution.

As before, this new equation can be solved by an iterative solution technique. However, Gortler and Cohen point out that in solving this equation it is better *not* to represent the matrix explicitly, as the new matrix (after preconditioning) is no longer as sparse as the original. Instead, they suggest an "implicit" approach, in which each step of the iterative solution method—which involves a matrix–vector multiply—is computed by running a filter-bank algorithm, as described in Section 7.1.2. This implicit approach allows each matrix–vector multiplication to be performed in a linear amount of time and space.

The effect of this change of basis, or preconditioning step, is dramatic, as illustrated in the lower row of Figure 12.2. Note that with the wavelet basis the curve converges to the correct solution much more quickly, achieving close to its final form after just 64 iterations. Such a basis is much more effective in supporting an interactive variational curve or surface design tool.

12.5 Adaptive variational modeling

Choosing an a priori fixed set of basis functions may not always be the best way to model curves or surfaces. If the set of basis functions is too small, the finite-element solution to the variational modeling problem may not be a good approximation of the optimal curve or surface. On the other hand, if the set of basis functions is too large, the solution procedure may become prohibitively expensive. Therefore, it would be ideal to start with a few coarse basis functions and gradually include finer basis functions where they are needed to improve the solution. The procedure used to decide which basis functions to add to the solution is sometimes called an *oracle*, as it predicts where the solution should be refined.

The oracle described by Gortler and Cohen [49] incorporates new basis functions into the solution in two situations. First, if a constraint is not currently being met, the oracle adds finer wavelets in the region of parameter space closest to that of the constraint. Second, the oracle includes finer wavelets near existing basis functions with coefficients above some threshold, where finer detail may help the solution attain a lower energy. The oracle can also "deactivate" basis functions whenever their coefficients drop below some minimum threshold. In this way, only those wavelets that contribute significantly to the optimal solution will be involved in the iterative solution process.

The combined efficiency of using a wavelet basis along with an algorithm that adapts the basis to a particular problem allows fairly difficult modeling tasks to be performed at interactive speeds. The images in Color Plate 13 illustrate an interactive surface editor. The user specifies the position of the surface in some places and the normal direction of the surface in others, and the application automatically finds the minimal-energy surface meeting those constraints.

Using an adaptive algorithm to selectively refine a wavelet-based solution turns out to be a handy technique for a wide variety of problems. In fact, we will see another example of this approach when we use a wavelet basis to solve global illumination problems in the next chapter.

GLOBAL ILLUMINATION

1. Radiosity — 2. Finite elements and radiosity — 3. Wavelet radiosity —
4. Enhancements to wavelet radiosity

Photorealistic image synthesis is one of the most fundamental problems in computer graphics, with applications to such diverse fields as scientific visualization, lighting and industrial design, remote sensing from satellites and robots, and entertainment and advertising.

The approach known as *global illumination* attempts to create realistic images by simulating the interreflection, emission, and absorption of light by surfaces in a scene. All of these interactions are governed by a set of physical laws that can be described by an integral equation. Unfortunately, this equation is complex enough for general scenes that approximation techniques must be employed.

In this chapter, we'll look at a mathematical description of the global illumination problem, and we'll see how wavelets can be used in its solution.

13.1   Radiosity

We will begin by looking at a special case of the global illumination problem, the case in which all surfaces are ideally *diffuse*—that is, they reflect and emit light equally in all directions. This special case is commonly known as the *radiosity problem* because it involves computing a solution to a transport equation that involves radiosity—a measure of energy per unit time and area. More formally, we can define *radiosity* $B(x)$ as the power per unit area that leaves a point x on a surface. Radiosity B is measured in [watt · meter^{-2}].

The equilibrium distribution of radiosity satisfies the following *radiosity transport equation* [22]:

$$B(y) = B_e(y) + \frac{\rho(y)}{\pi} \int_x G(x, y) B(x) \, dx$$

(13.1)

This equation states that the radiosity B at a point y is the sum of two terms: *emitted radiosity* B_e and the integral of radiosity reflected from all other points x. An infinitesimal area around point x is written dx. The term $\rho(y)$, the *reflectance*, describes the fraction of incident energy that is reflected rather than transmitted through the surface or absorbed (and converted to heat). Since it is a ratio, the reflectance is dimensionless and ranges from 0 to 1. The π in the equation is a normalization factor that accounts for the integration over the hemisphere of directions in which light reflects from y. Finally, the *geometric term* $G(x, y)$ describes how radiosity leaving a differential area at x arrives at y. It is given by

$$G(x, y) := V(x, y) \cdot \frac{\cos \theta_x \cos \theta_y}{\|x - y\|^2}$$

where $V(x, y)$ is a *visibility term* that is 1 or 0, depending on whether or not x and y are visible to one another, and θ_x and θ_y are the angles between the line segment xy and the respective normals of differential areas at x and y. The geometric term is symmetric in its arguments: $G(x, y) = G(y, x)$. Some of these terms are illustrated in Figure 13.1.

It is often convenient to be able to suppress most of the detail in the radiosity transport equation equation (13.1) and rewrite in the simpler-looking *operator form*:

$$B = B_e + \mathcal{T}B$$

(13.2)

Here, the *radiosity transport operator* $\mathcal{T}$ is defined by

$$(\mathcal{T}B)(y) := \frac{\rho(y)}{\pi} \int_x G(x, y) B(x) \, dx$$

where $(\mathcal{T}B)(y)$ denotes the result of $\mathcal{T}$ operating on $B(x)$ to produce a function whose argument is y. Intuitively, you can think of the transport operator $\mathcal{T}$ as performing one "bounce" in

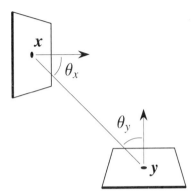

FIGURE 13.1   The geometry involved in radiosity transport from point x to point y.

the transport of light: for every surface (parameterized by y), it gathers radiosity from all other surfaces in the scene (parameterized by x) and reflects it back out into the scene according to the reflectance $\rho(y)$.

Note that up until now we have not mentioned color in our presentation of the radiosity transport equation. In the absence of fluorescence (the phenomenon of a material absorbing light of one wavelength and re-emitting it at another), we can simply make B, B_e, and ρ all functions of wavelength as well as location. Then equation (13.2) holds for any particular wavelength we care to consider.

13.2   Finite elements and radiosity

To determine the exact solution to the radiosity problem for a particular scene, we would have to find the amount of light leaving every point on every surface. That is, we would have to solve the radiosity transport equation—equation (13.1) or equation (13.2)—for the radiosity $B(y)$ for all points y.

Unfortunately, in both forms of the equation, the unknown function B appears both inside and outside an integral. There are very few situations in which the geometry, reflectivity, and emission terms allow an analytical solution to this equation to be found. We therefore turn to the finite-element approach to compute an approximate solution. As we did in Chapter 12, we will rewrite the unknown function as a linear combination of basis functions. As a result, the problem is cast as a linear system of equations in a set of unknown coefficients—something a computer can solve. But before we arrive at this linear system, we need to go into more detail about how basis functions can be used to discretize radiosity.

13.2.1 Discretizing radiosity

Let $u(x) = [u_1(x) \quad u_2(x) \quad \cdots]$ be a basis for the space of radiosity functions. The unknown radiosity distribution B can be expressed as a linear combination of the basis functions $u_i(x)$ with unknown coefficients b_i:

$$B(x) = \sum_{i=1}^{\infty} b_i\, u_i(x)$$

This equation can be written in matrix form as $B(x) = u(x)B$, where B is an infinite column matrix whose i-th entry is b_i. When no confusion can arise, we suppress the arguments and simply write

$$B = uB$$

13.2.2 Discretizing radiosity transport

Once we've written radiosity in terms of basis functions, we can obtain a system of equations for the unknown entries of B. We first substitute $B = uB$ and $B_e = uB_e$ into equation (13.2), and then use linearity of the operator $\mathcal{T}$ to yield

$$uB = uB_e + \mathcal{T}(uB) = uB_e + (\mathcal{T}u)B \tag{13.3}$$

The next step is to project equation (13.3) back onto the basis u. For this step, we need to use the *dual basis* associated with u, which we denote by $\tilde{u} = [\tilde{u}_1(x) \quad \tilde{u}_2(x) \quad \cdots]$. As was discussed in Section 7.4.1, the dual basis is characterized by the relation $\langle \tilde{u}_i \mid u_j \rangle = \delta_{ij}$, or in matrix form $[\langle \tilde{u} \mid u \rangle] = I$, where I is the identity matrix. Recall that orthonormal bases such as the Haar basis are *self-dual*, meaning that $\tilde{u} = u$.

We can project equation (13.3) onto our basis by taking the inner product of both sides of the equation with each dual basis function in turn. In matrix notation, this projection is accomplished by applying the linear operator $[\langle \tilde{u} \mid \cdot \rangle]$ to both sides of the equation to get

$$[\langle \tilde{u} \mid uB \rangle] = [\langle \tilde{u} \mid uB_e \rangle] + [\langle \tilde{u} \mid (\mathcal{T}u)B \rangle]$$

Using linearity, we can factor the matrix $[\langle \tilde{u} \mid u \rangle]$ out of the left-hand side and the first term of the right-hand side. According to the duality relation, this matrix is actually the identity, and so we can remove it altogether. We thereby arrive at the *discrete radiosity transport equation:*

$$B = B_e + TB \tag{13.4}$$

In this infinite system of linear equations, the *transport matrix* $T := [\langle \tilde{u} \mid \mathbb{T}u \rangle]$ is an infinite matrix representing the transport operator $\mathbb{T}$. The (r, s)-th entry of T is a *transport coefficient*, representing the influence of the coefficient of u_s on the coefficient of u_r. It can be written explicitly as

$$T_{r \leftarrow s} = \langle \tilde{u}_r \mid \mathbb{T}u_s \rangle$$

$$= \left\langle \tilde{u}_r \,\middle|\, \frac{\rho(y)}{\pi} \int_x G(x, y)\, u_s(x)\, dx \right\rangle$$

$$= \int_y \tilde{u}_r(y)\, \frac{\rho(y)}{\pi} \int_x G(x, y)\, u_s(x)\, dx\, dy \tag{13.5}$$

where the notation $r \leftarrow s$ serves to emphasize that $T_{r \leftarrow s}$ represents the influence of the *sender s* on the *receiver r*. In the integral above, the domain of x is the support of the sending basis function u_s, and the domain of y is the support of the receiving basis function $\tilde{u}_r$.

13.2.3 Solving the linear system

The discrete version of the radiosity transport equation, equation (13.4), still represents an infinite set of equations and unknowns. The most straightforward way to find an approximate solution to this set of equations might be to choose a priori a large, fixed set of basis functions with which to represent the solution. For example, we could split all surfaces of the input scene into m surface patches and then erect a single box function over each of these patches to serve as our basis. We could then compute all m^2 entries of T according to equation (13.5) (a task that requires nontrivial numerical or analytical integration). Finally, we could use any linear equation solver to find the m unknown radiosity coefficients that satisfy equation (13.4).

However, there is a very serious drawback to such an approach: the cost of computing the transport matrix T far outweighs the cost of solving the linear system that results. The principle cost is in evaluating equation (13.5) for each of the pairs of sending and receiving basis functions. A far better approach is to first compute just a few of the entries of T, then find the corresponding radiosity solution B, and only then add more entries to T, based on what we've found so far.

The algorithm known as *hierarchical radiosity* [55] takes just such an approach. It first solves for the average radiosity of each surface in the scene, then splits each rectangular region in need of refinement into four subregions, and then solves again. It repeats this process of alternately refining and solving until the solution has converged sufficiently. The basis functions underlying hierarchical radiosity are piecewise-constant box functions of varying sizes. The wavelet-based method we'll present next is similar, but instead of using box functions, we will use Haar wavelets to represent the detail in the refined solution that is not present in the initial coarse solution.

13.3   Wavelet radiosity

Hierarchical or multiresolution techniques for solving global illumination problems all follow the same basic steps: they iteratively refine the interactions between the basis functions of an existing solution and then use these interactions to compute an improved solution. Multiresolution methods exploit the fact that in some parts of the scene the solution can be represented with sufficient accuracy using only a few basis functions. Even where many basis functions are needed, each basis function typically interacts with only a few others.

Before we present a wavelet-based radiosity algorithm, we need to consider how radiosity and the transport operator will be represented using wavelets.

13.3.1   A wavelet basis for radiosity

Radiosity is defined on a two-dimensional domain, namely, the surfaces in the scene. We'll assume that the surfaces are split into regions that can be parameterized on the unit square. Therefore, we need two-dimensional basis functions defined on the unit square. As we pointed out when dealing with images in Chapter 3, the simplest way to develop a multidimensional wavelet basis is to take tensor products of one-dimensional basis functions. Thus, we need to decide which one-dimensional wavelet basis to use and whether to use the standard construction or the nonstandard construction to form a two-dimensional basis.

For the sake of simplicity, we'll use the Haar basis in our explanation. Other bases that have been used in wavelet radiosity algorithms include *flatlets* and *multiwavelets* [50, 107]. We'll use the nonstandard construction of two-dimensional basis functions because it allows us to store a sparse set of wavelet coefficients very easily in a quadtree, just as we did for image editing in Chapter 4.

13.3.2 Data structures for wavelet radiosity

The radiosity of each patch in a scene can be thought of as an image. If we are interested in radiosity solutions at wavelengths roughly corresponding to red, green, and blue, we can use a quadtree data structure that is very similar to the one we used for image editing in Chapter 4:

> **type** *QuadTree* = **record**
>   *c*: *RGB*
>   *root*: **pointer to** *QuadTreeNode*
>   *links*: **pointer to** *Link*
> **end record**

> **type** *QuadTreeNode* = **record**
> d_1, d_2, d_3: *RGB*
>   *child*: **array** [1 .. 4] **of pointer to** *QuadTreeNode*
>   *links*: **pointer to** *Link*
> **end record**

By analogy to the image case, *c* represents the overall average radiosity of a patch, while the d_i coefficients multiply the wavelets $\psi\phi$, $\phi\psi$, and $\psi\psi$ at each level and translate of the quadtree. The *child* array contains pointers to the quadtree nodes that store finer levels of detail in each of four quadrants.

Each structure also contains a field called *links*, which points to a dynamic list of transport coefficients $T_{r \leftarrow s}$ from all other sending nodes *s* that have this node *r* as a receiver. Here is the data structure for a link:

> **type** *Link* = **record**
>   *s*: **pointer to** *QuadTree* **or** *QuadTreeNode*
>   *T*: **array** [1 .. 9] **of** *RGB*
>   *variation*: **real**
>   *next*: **pointer to** *Link*
> **end record**

The *Link* structure can contain up to nine transport coefficients, which allows it to represent the transport coefficients for all combinations of wavelets on the sender and receiver. A link between a *QuadTree* structure (containing a scaling function coefficient) and a *QuadTreeNode* structure (containing three wavelet coefficients) uses just three of these transport coefficients, while a link between two *QuadTree* structures uses just one.

13.3.3   Wavelet radiosity algorithm

Now we're equipped to describe a wavelet radiosity algorithm. Our primary task is to solve a system of linear equations for radiosity:

$$B = B_e + TB$$

We first compute a small number of entries of the matrix T and solve the equations, then compute more entries of T and solve again, and so on. Because the entries of T are very expensive to compute, we strive to compute as few of these entries as possible while generating a good approximation to the solution. Put briefly, only entries of T that are estimated to be large—and that connect large basis function coefficients—are computed.

The main part of the algorithm alternates between computing an approximate radiosity solution $\hat{B}$ and improving the finite representation of the transport operator $\hat{T}$. Quantities with a hat are approximate, both because they are computed numerically and because they are truncated versions of infinite matrices. Initially, we project B_e into space V^0, the space spanned by the coarsest-level scaling functions, to give $\hat{B}_e$. We also compute the entries of T corresponding to interactions of scaling functions in V^0 with one another (as described in Section 13.2.2), giving $\hat{T}$. The algorithm is given in pseudocode below:

> **procedure** *GlobalIllumination*($\hat{T}$, $\hat{B}_e$)
> $\hat{B} \leftarrow \hat{B}_e$
>   **repeat**
> $\hat{B} \leftarrow$ *Solve*($\hat{T}$, $\hat{B}$, $\hat{B}_e$)
> $\hat{T} \leftarrow$ *Refine*($\hat{T}$, $\hat{B}$)
>   **until** convergence of $\hat{B}$
> **end procedure**

The main loop iterates until convergence is achieved, that is, until further refinement does not significantly change the computed radiosity solution. The *Solve* function uses Gauss-Seidel iteration [47] to solve the approximate transport equation $\hat{B} = \hat{B}_e + \hat{T}\hat{B}$. The *Refine* function is described in the next section.

13.3.4   Refinement

The goal of refinement is to determine which of the entries of T missing from $\hat{T}$ should be computed to reduce the error in the current solution. The function we will describe here uses concepts from the brightness-weighted refinement criterion for hierarchical radiosity [55] and the refinement oracle used by Gortler et al. for wavelet radiosity [50]. The idea is to estimate the energy transfer that would take place if a new transport coefficient were to be included in $\hat{T}$. If this quantity falls below some threshold, the expensive computation of the transport coefficient can be avoided without resulting in significant error in the solution.

There are infinitely many new transport coefficients that we could incorporate into $\hat{T}$. We need a scheme that allows us to consider only a few of them in each iteration while making it possible to eventually consider them all.

A simple approach is to take each of the existing transport coefficients and consider refining the basis functions at the sending end or the receiving end or both. The transport coefficients that would result from each of these possible refinements are illustrated in Figure 13.2. Refinement at the sending end, shown in Figure 13.2(a), creates transport coefficients linking all the children of the sending basis function to the receiving basis function. Refinement at the receiving end, shown in Figure 13.2(b), creates transport coefficients linking the sending basis function to all the children of the receiving basis function.

Suppose we are considering two basis functions, u_s and u_r, that are not yet connected by a transport coefficient. We will compute a new transport coefficient $\hat{T}_{r \leftarrow s}$ if a sufficiently large value results from the product of

- *radiosity:* the magnitude of the sending basis function coefficient $\hat{b}_s$ and

- *estimated transport coefficient*: the estimated new transport coefficient $\hat{T}_{r \leftarrow s}$ between the basis functions

The sending basis function coefficient is easily computed through filter-bank reconstruction. However, we also need a way of estimating the transport coefficient without going to the expense of actually computing it. What we really need to know is the extent to which the kernel deviates from a constant function, because if it is nearly constant over the support of the sending and receiving basis functions, we do not need any more wavelets. One measure of how far the kernel deviates from a constant function is its *variation*—the maximum value minus the minimum value (over all three color channels). Thus, if each link stores a transport coefficient as well as the variation in the kernel samples used to compute that transport coefficient, we can later use the variation when we consider refining that link.

The oracle described by Gortler et al. [50] uses a similar approach to refinement. In order to take advantage of wavelets with more vanishing moments than the Haar basis, their algorithm refines when the kernel is not well approximated by an interpolating polynomial of the appropriate degree.

As mentioned earlier, each *Link* structure actually stores up to nine transport coefficients in order to accommodate all combinations of sender and receiver basis functions between any

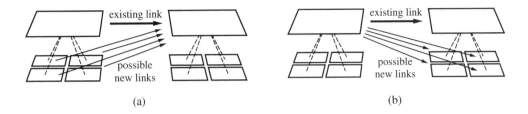

FIGURE 13.2   Links to be considered when refining an existing link between basis functions: (a) refinement at the sending end; (b) refinement at the receiving end.

two nodes. Note that for all nine transport coefficients, both the estimated radiosity of the source and the kernel variation will be the same. Thus, the estimated contribution of all nine transport coefficients will also be identical, and it makes sense, therefore, to compute all of these transport coefficients at the same time.

One final detail of the refinement process involves the choice of the threshold value above which new links are created. One approach is to begin the first refinement with some arbitrary constant for the threshold and proceed by cutting the threshold in half each time the *Refine* function is called. Alternatively, we can compute the maximum energy transfer $\hat{T}_{r \leftarrow s} \hat{b}_s$ of the existing links and take some fixed fraction (say, 90%) of that value as the threshold.

13.3.5   Transport matrix decompositions

The algorithm we have just presented is based on what is called a *standard decomposition of the transport operator*. In other words, just as there are two different wavelet transforms for images and other functions over a two-dimensional domain (the standard and the nonstandard), so there are also two ways to decompose the two-dimensional matrix that represents the transport operator.

Consider what happens when we use a large number of very fine box functions as our basis for representing radiosity. When we discretize a global illumination problem using this box function basis, we end up with a large, dense transport matrix T. Now, in order to arrive at a sparse transport matrix that represents the same information, we would like to take the wavelet transform of T. By analogy with image compression, we can expect the wavelet transform to produce a matrix with many insignificant entries, and we can formulate that wavelet transform in two different ways.

The approach we have described for wavelet radiosity corresponds to the standard decomposition. Our discretization of the transport operator allowed every basis function to interact with every other one directly, regardless of level. The nonstandard decomposition, on the other hand, only allows a given wavelet to interact with other wavelets at a particular level in the hierarchy.

In this text, we won't go into any more detail about wavelet radiosity algorithms based on the nonstandard decomposition of the transport operator. Instead, we refer the interested reader to the articles by Gortler et al. [50], Schröder et al. [107], and Schröder and Hanrahan [108]. One potentially important consideration is that if n is the number of basis functions, theory predicts that for operators with certain smoothness constraints a nonstandard decomposition will have $O(n)$ significant entries, and a standard decomposition will have $O(n \log n)$ [5]. However, for radiosity transport, the nonstandard decomposition often appears to be no more sparse then the standard decomposition in practice [12].

13.3.6 Final gathering step

Up to this point, we haven't mentioned how a solution to the global illumination problem is used to create an image. Suppose we have a radiosity solution in the Haar wavelet basis and we wish to render it from a particular viewpoint. One technique is to apply wavelet synthesis to reconstruct a scaling-function representation of the radiosity on each patch and then draw each of the small regions of constant radiosity as a constant-colored rectangle. Another possibility is to cast rays from the viewpoint through each pixel of the image plane and evaluate the radiosity solution at the first surface location that each ray hits. Both of these rendering methods display all of the discontinuities that are inherent in a piecewise-constant representation such as the Haar basis.

One way to avoid some of the artifacts of a piecewise-constant basis is to perform a *final gathering step* that recomputes the illumination arriving at each surface point corresponding to a pixel in the image [71, 100, 117]. For each pixel in the image, we cast a ray to find the surface corresponding to that pixel. Then we gather light from all the basis functions that contribute to the radiosity solution at that point. In other words, for each visible surface point y we compute

$$B_e(y) + \frac{\rho(y)}{\pi} \sum_s \int_x G(x, y) \, B_s(x) \, dx$$

where the sum over sending patches s includes all patches that have basis functions linked to receiving basis functions at the point y.

Formally, the final gather corresponds to reprojecting the solution onto a new basis, where each basis function is a delta function located at a visible point. This new basis is tailored for visual accuracy. The final gather smooths the discontinuities in the wavelet representation and makes textures and shadows crisper, as demonstrated by the difference between the images in Color Plates 14(b) and 14(c).

13.4   Enhancements to wavelet radiosity

The algorithm outlined in the previous section can be extended in a number of ways to improve its speed, efficiency, and quality of solution. We'll describe a few of these possible enhancements.

13.4.1   Importance-driven refinement

Global illumination algorithms are typically *view independent*, meaning that the solution, once computed, can be used to render images of the scene from any viewpoint. While this view independence is advantageous for some applications, such as an architectural walkthrough, in other applications we may only care about a single view or a very limited set of views. In this case, we could waste a lot of time computing an accurate radiosity solution everywhere when only a very small portion of the scene might ever actually become visible.

Smits et al. [119] describe an extension to hierarchical radiosity that uses *importance functions* (originally from neutron transport theory) to describe how much the radiosity solution at a given location influences the visible part of the scene. The technique they describe is directly applicable to wavelet radiosity as well. The basic idea is to define a dimensionless quantity called importance that is transported from surface to surface just like radiosity but that is emitted from the viewpoint rather than from light sources. Importance can then be included in the refinement criterion so that only interactions that are both bright and important will be refined.

13.4.2   Clustering

Clustering is another extension to hierarchical radiosity that applies equally well to wavelet radiosity. Both hierarchical and wavelet radiosity require that we initially compute transport coefficients between all pairs of input patches. If n is the number of input patches, then for very complex scenes, just creating the $O(n^2)$ initial links may be incredibly time-consuming. Clustering algorithms attempt to reduce this cost by grouping together nearby patches and by starting with coarse interactions between these clusters.

In one such approach, Smits et al. [118] show how clusters can be created automatically, using error bounds to guide the solution process. Their algorithm has just $O(n \log n)$ complexity. A similar approach can be used in a wavelet-based algorithm, as shown by Christensen et al. [13].

13.4.3 Generalization to radiance

When we attempt to solve the global illumination problem using the radiosity transport equation, we are dealing with a special case in which all surfaces in the environment are ideally diffuse. While this assumption is a reasonable one for certain environments, such as interiors covered with latex paint, it is overly restrictive for many other situations that commonly arise. It only takes a single glossy surface in a scene to violate the assumptions of radiosity.

The more general form of the global illumination problem, known as glossy global illumination, has considerably higher complexity. Whereas radiosity is concerned with how light reflecting from every surface point affects light reflecting from all other points, in glossy global illumination one must consider how light reflecting *in every direction* from every surface point affects light reflecting *in all other directions* from all other surface points. Instead of solving for radiosity, we are now interested in solving for *radiance*, which is a function of both position and direction.

Radiance can be defined formally as a function $L(x, \omega)$ representing the power emanating from a surface point x per unit solid angle in the direction ω per unit projected area perpendicular to that direction. Radiance L is measured in [watt $\cdot$ meter^{-2} $\cdot$ steradian^{-1}]. (A *steradian* is a measure of "solid angle"; just as there are 2π radians in a circle, there are 4π steradians in a sphere.) The equilibrium distribution of radiance satisfies the following *radiance transport equation* [22, 114]:

$$L(y, \omega) = L_e(y, \omega) + \int_x f_r(\omega_{xy}, y, \omega) \ G(x, y) \ L(x, \omega_{xy}) \, dx$$

Here ω_{xy} is the direction from x to y and $f_r(\omega_{xy}, x, \omega)$ is the *bidirectional reflectance distribution function*, or BRDF. The BRDF is measured in [steradian^{-1}] and describes the ratio of reflected radiance (in direction ω) to the differential irradiance (from direction ω_{xy}) that causes it. The geometric term $G(x, y)$ is the same as it was for radiosity.

Clearly, the most significant change when going from wavelet radiosity to wavelet radiance is that the solution is a function of four variables: two describing position on a surface and two describing direction. While surface positions on a patch can easily be parameterized on the unit square $[0, 1]^2$, the set of directions over a surface point are most naturally parameterized on the unit hemisphere H^2. One approach we could take is to define a wavelet basis for the domain $[0, 1]^2 \times H^2$, for which we might make use of the spherical wavelets described by

Schröder and Sweldens [109]. An alternative approach was taken by Christensen et al. [15], who mapped the directions on the hemisphere to points on the unit square, making it possible to construct a wavelet basis for radiance using the domain $[0, 1]^4$.

Aside from the higher dimensionality of radiance, a wavelet-based algorithm for radiance is very similar to the one we described for radiosity [15]. The images in Color Plate 15 illustrate the kinds of effects that can be achieved in this more general characterization of the global illumination problem.

14

FURTHER READING

1. Theory of multiresolution analysis — 2. Image applications — 3.Curve and surface applications — 4. Physical simulation

We have only addressed a small fraction of the material that's currently available on wavelets and multiresolution techniques. In this chapter, we briefly survey some additional material.

14.1   Theory of multiresolution analysis

Our focus in this text has been on the shift-variant form of multiresolution analysis, as this setting more naturally accommodates the kinds of data sets that arise in computer graphics applications, such as curves defined on bounded intervals and surfaces defined on domains of arbitrary topological type.

The vast majority of the existing literature on wavelets, however, assumes shift invariance. The general shift-invariant theory is beautifully described in the book by Daubechies [25]. The book by Chui [17] also contains a very nice introduction to shift-invariant multiresolution analysis, and it presents a thorough treatment of wavelets constructed from uniform B-splines. The generalization to shift-invariant wavelets based on multivariate B-splines can be accomplished with tensor-product B-splines or, more generally, with box splines. The generalization via tensor products is straightforward, resulting in either standard or nonstandard constructions. The generalization using box splines is considerably more involved and is covered by Jia and Micchelli [64].

In the shift-variant setting, the simplest situation is the construction of wavelets for a bounded interval. Meyer [81] presents a method for adapting Daubechies scaling functions and wavelets to a bounded interval by introducing new functions near the boundaries. Chui and Quak [18] make extensive use of the theory of nonuniform B-splines to construct bounded-interval B-spline wavelets. In contrast to Section 7.3.1, in which an algorithmic construction was given, Chui and Quak give explicit formulas for minimally supported wavelets of arbitrary order. The results of their paper are sufficient for performing linear-time filter-bank synthesis. The development of a linear-time analysis algorithm was later presented by Quak and Weyrich [98]. Like the analysis algorithm described in Section 7.3.1, Quak and Weyrich's algorithm is based on solving a banded linear system, although the system they solve is somewhat different.

The general shift-variant setting based on subdivision was first introduced by Lounsbery et al. [75]. It has since been expanded upon by Sweldens [123], who recognized the wavelet construction of Lounsbery et al. as one instance of the more general idea of lifting. Sweldens also showed that in the univariate case all biorthogonal schemes are related by lifting; that is, there is a lifting operation that turns any biorthogonal scheme into any other biorthogonal scheme. Schröder and Sweldens [109] have continued this work by demonstrating that subdivision and lifting provide constructive methods for building custom-crafted wavelets, an approach they refer to as *second-generation wavelets*. They concentrate on wavelet representations for functions defined on the sphere, where they take into account the curved geometry of the sphere in addition to its topological type. (Our treatment in Chapter 10, by contrast, assumes the domain, or "base mesh," to be a piecewise-linear surface.) Schröder and Sweldens show that using the geometry of the sphere as the domain results in improved compression rates.

Since shift-variant multiresolution analysis is based on subdivision, the development of a general theory for the analysis and construction of subdivision schemes is an important and related problem. Although a complete theory does not yet exist, quite a lot is known about various special cases. There is a very extensive and elegant theory, for instance, for stationary uniform subdivision schemes, developed largely by Dahmen, Dyn, Gregory, Levin, and Micchelli. An excellent (though advanced) treatment is available in the book by Cavaretta et al. [8]. Advances in the stationary nonuniform case have recently been obtained by Reif [101] and Warren [127], but much is still unknown. Even less is known about nonstationary subdivision. See the paper by Dyn and Levin [31] for a relatively recent survey of the state of the art in subdivision.

14.2 Image applications

There is a huge body of literature devoted to the use of wavelets in image compression and image processing; the first part of this book has only scraped the surface. The image compression algorithms discussed in Chapter 3 are fairly rudimentary. They can be improved by using other wavelet bases or by using transformations that adapt better to the image data than do the standard or nonstandard wavelet transforms. The *wavelet packet transform* [133] is one such method. In the usual wavelet transform, as described in Chapter 2, we recursively decompose the scaling function coefficients into two sets, leaving the wavelet coefficients at each level alone. By contrast, the wavelet packet transform allows us to decompose both the scaling function coefficients and the wavelet coefficients at each level, using the same analysis filters *A* and *B*. The decision about whether or not to decompose a set of coefficients is made based on which choice provides the greatest opportunity for compression. For more on compressing images using wavelet packets, see the work done by Wickerhauser [132, 133].

One of the drawbacks of the Haar basis, particularly for image querying, is its sensitivity to translations—the wavelet decomposition of an image changes fairly dramatically if the image is translated slightly. Even wavelet bases that are constructed to have better properties than the Haar basis often suffer from this same instability with respect to translation, rotation, or scaling. A special class of wavelet transforms called *shiftable transforms* or *steerable filters* [115] is designed to deal robustly with these geometric transformations. As we mentioned in Chapter 5, shiftable transforms might enable a multiresolution image querying algorithm to behave more robustly when given translated, rotated, and scaled queries. In the related application of optical character recognition, Shustorovich [113] uses shiftable Gabor wavelets to find edges in images. A different approach to edge detection is presented by Mallat and Zhong [79]; they describe how the maxima in a continuous wavelet transform can be used to locate edges in an image and, furthermore, how the image can be reconstructed from this edge information alone.

In a good deal of our discussion of images we have assumed that we are given an image as input. However, we sometimes want to generate an image from scratch that has certain characteristics. The problem of *texture generation* is particularly amenable to wavelet techniques. Perlin and Velho [94] describe one way of creating a texture with infinitely fine scales of detail within the context of their multiresolution painting system. Their idea is to define a procedure capable of algorithmically generating detail coefficients corresponding to some pattern as a user magnifies an image. A related approach to generating textures is to take the wavelet transform of an existing image and examine its statistical properties, then create a new multiresolution image with random numbers having the same statistical distribution. Heeger and Bergen [58] use this approach to generate large periodic textures from smaller, nonperiodic scanned images.

Heeger and Bergen also describe how the statistics of a single image (a photograph of marble, for instance) can be used to generate a three-dimensional *solid texture* (for example, a block of marble) with the same statistical properties in each slice. In fact, most multiresolution image algorithms can easily be extended to three dimensions. Wavelet compression algorithms can be applied to volumetric data, and in fact, the resulting representation can be used to accelerate volume rendering [88, 89, 131]. Wavelet transforms of volume data have also been used to accelerate and improve on previous volume-to-volume morphing techniques [56].

Video sequences are also three-dimensional, although the temporal dimension is somewhat different from the two spatial dimensions of each frame in a video. This difference distinguishes the compression of video from the compression of volume data. It is possible, and usually desirable, to reconstruct a frame in a compressed video sequence without knowing all of the future frames. Lewis and Knowles [70] describe an approach to video compression in which a wavelet transform is applied to each frame, and then each group of eight frames is transformed in the time dimension to take advantage of temporal coherence. Another approach is to take the wavelet transform of each frame and use frame differences or motion estimation techniques to predict the next frame from the current one [91, 112, 136]. Compression and progressive transmission of video sequences are just two of the many possible applications of a multiresolution representation for video—Finkelstein et al. [36] describe interesting ways of exploring and editing multiresolution video.

14.3   Curve and surface applications

There are a number of applications beyond the ones presented in Part II that have benefited from multiresolution representations of curves. For instance, Reissell has constructed interpolating "pseudocoiflets" for use in path planning and curve intersection testing [26]. Wavelets also figure prominently in the work by Chun and Kuo [19], in which wavelets are used to improve the two-dimensional shape blending method originally introduced by Sederberg et al. [110, 111]. Witkin and Popović [135] present a technique for "motion warping" of animation trajectories that does not require wavelets but might benefit from their use.

In the realm of surface applications, it is worth noting the pioneering work of Forsey and Bartels [39] on hierarchical B-splines. Although their construction does not use wavelets per se, their work is the first instance of multiresolution surface definition and editing. Forsey and Bartels [40] have since used hierarchical B-splines to solve surface-fitting problems. Concise fitting of surfaces has also been addressed by DeVore et al. [28], who use quartic box-spline wavelets. Among the interesting aspects of this paper are the bounds on the L^∞ error introduced by their compression procedure.

14.4 Physical simulation

Global illumination and variational modeling are both examples of *physical simulation* problems. In variational modeling, the "quality" of a curve or surface is measured by the physical energy of a bent wire or plate. In global illumination, the radiosity transport equation itself is derived from the physics of radiative transfer. As discussed in the previous two chapters, these problems can both be approached by discretizing the unknown solution function and representing it with finite elements, such as wavelets in particular.

A good overview of variational surface modeling is given in Welch's thesis [129]. In other work related to variational modeling, Halstead et al. [54] have presented a method for designing smooth interpolating surfaces of arbitrary topology using Catmull-Clark subdivision while minimizing a "fairness" norm. By using k-disc wavelets based on the butterfly scheme (Chapter 10), their work could probably be extended to a multiresolution approach. We should mention that hierarchical bases other than wavelets have been used to precondition and accelerate algorithms for surface fitting (see Szeliski's paper [124], for example). In addition, Pentland [92] has shown how wavelet-based methods for surface fitting can be used to solve a wide variety of physical equilibrium problems.

A great deal has been written about the global illumination problem, including good introductions to the subject by Cohen and Wallace [22], Sillion and Puech [114], and Glassner [46]. The development of multiresolution methods for radiosity begins with the hierarchical radiosity formulation presented by Hanrahan et al. [55] and continues with the wavelet-based work by Gortler, Schröder, and colleagues [50, 107]. Many of the further enhancements to wavelet radiosity algorithms that were mentioned in Chapter 13 are discussed in detail in Christensen's thesis [12] and in his joint work with the authors of this book [13, 14, 15]. The spherical wavelets developed by Schröder and Sweldens [109] are obvious candidates for representing the angular distributions of radiance that are required to solve global illumination problems in the presence of glossy surfaces.

Many other physical simulation problems can be solved with finite elements and may also benefit from a multiresolution approach. For example, radiative heat transfer, acoustics simulations, electrostatics problems, and the n-body gravitational problem all bear resemblance to the physical simulation problems described in the previous two chapters.

One area of particular interest in computer graphics, which deserves special mention, is animation. The goal of realistic animation is to portray a figure accomplishing some task while obeying physical laws and expending a minimum of energy. A mathematical formulation of this goal function results in a variational problem in which the objective is to minimize an integral representing the total work, and the constraints are the laws of physics as well as any initial and final conditions imposed by the animator. This framework for automatically producing realistic-looking animations, called *spacetime constraints*, was introduced by

Witkin and Kass [134] and later refined by Cohen [21]. Their implementations use fixed-resolution bases to convert a variational problem to a nonlinear optimization problem. As with the variational modeling problem, a fixed-resolution approach suffers from slow convergence and poor conditioning. Liu et al. [72] show that a B-spline wavelet basis dramatically improves the rate of convergence and allows the optimization process to adapt to details of the animation that are localized in time.

APPENDICES

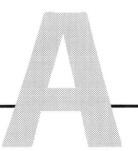

LINEAR ALGEBRA REVIEW

1. Vector spaces — 2. Bases and dimension — 3. Inner products and orthogonality — 4. Norms and normalization — 5. Eigenvectors and eigenvalues

Recursive subdivision and multiresolution analysis rely heavily on fundamental ideas from linear algebra. We've included this appendix to review some of the important concepts of linear algebra.

A.1   Vector spaces

The starting point for linear algebra is the notion of a *vector space*. Suppose V is a collection of "things" for which addition and scalar multiplication are defined. Then the set V is a vector space (over the reals) if the following axioms hold for any $u, v, w \in V$ and $a, b \in \mathbb{R}$:

1. $u + v = v + u$

2. $(u + v) + w = u + (v + w)$

3. There is an element $\mathbf{0} \in V$ such that $\mathbf{0} + u = u$

4. For each $u \in V$, there is an element $-u \in V$ such that $u + (-u) = \mathbf{0}$

5. $(a + b)u = au + bu$

6. $a(u + v) = au + av$

7. $(ab)u = a(bu)$

8. $1 \cdot u = u$

The elements of a vector space V are called *vectors*, and the element 0 is called the zero vector. The vectors may be column matrices of scalars like $[1 \ 0 \ 0]^{\text{T}}$, or they may be functions like the scaling functions $\phi(x)$ and wavelets $\psi(x)$ of multiresolution analysis.

Note that in this appendix, as in the rest of the text, we'll use boldface italics for vectors that are column matrices of scalars and normal italics for other types of vectors or when referring to vectors in general.

A.2   Bases and dimension

A collection of vectors $u_1, u_2, \ldots$ in a vector space V are said to be *linearly independent* if

$$c_1 u_1 + c_2 u_2 + \cdots = 0 \quad \text{if and only if} \quad c_1 = c_2 = \cdots = 0$$

A collection $u_1, u_2, \ldots \in V$ of linearly independent vectors is a *basis* for V if every $v \in V$ can be written as

$$v = \sum_i c_i u_i$$

for some real numbers $c_1, c_2, \ldots$. The vectors in a basis for V are said to *span* V. Intuitively speaking, linear independence means that the vectors are not redundant, and a basis consists of a minimal complete set of vectors.

If a basis for V has a finite number of elements $u_1, \ldots, u_m$, then V is *finite-dimensional* and its dimension is m. Otherwise, V is said to be *infinite-dimensional*. In this appendix, as in the rest of the text, we will use boldface to denote the vectors of finite-dimensional spaces

Example: $\mathbb{R}^3$ is a three-dimensional space, and $e_1 = [1 \ 0 \ 0]^{\text{T}}$, $e_2 = [0 \ 1 \ 0]^{\text{T}}$, $e_3 = [0 \ 0 \ 1]^{\text{T}}$ is a basis for it. ∎

Example: The set of all functions continuous on $[0, 1]$ is an infinite-dimensional vector space. ∎

A.3 Inner products and orthogonality

When dealing with geometric vectors from the vector space $\mathbb{R}^3$, the "dot product" operation has a number of uses. The generalization of the dot product to arbitrary vector spaces is called an *inner product*. Formally, an inner product $\langle \cdot \mid \cdot \rangle$ on a vector space V is any map from $V \times V$ to $\mathbb{R}$ that is

- symmetric: $\langle u \mid v \rangle = \langle v \mid u \rangle$

- bilinear: $\langle au + bv \mid w \rangle = a \langle u \mid w \rangle + b \langle v \mid w \rangle$

- positive definite: $\langle u \mid u \rangle > 0$ for all $u \neq \mathbf{0}$

A vector space together with an inner product is called, not surprisingly, an *inner product space*.

Example: It is straightforward to show that the dot product on $\mathbb{R}^3$ defined by

$$ \mathbf{a} \cdot \mathbf{b} := a_1 b_1 + a_2 b_2 + a_3 b_3 \tag{A.1} $$

satisfies the requirements of an inner product. ∎

Example: The following "standard" inner product on functions of the unit interval $[0, 1]$ plays a central role in most formulations of multiresolution analysis:

$$ \langle f \mid g \rangle := \int_0^1 f(x)\, g(x)\, dx $$

The standard inner product can also be generalized to include a positive weight function $w(x)$:

$$ \langle f \mid g \rangle := \int_0^1 w(x)\, f(x)\, g(x)\, dx $$

∎

One of the most important uses of the inner product is to formalize the idea of orthogonality: Two vectors u, v in an inner product space are said to be *orthogonal* if $\langle u \mid v \rangle = 0$. It is not difficult to show that a collection $u_1, u_2, \ldots$ of mutually orthogonal vectors must be linearly independent, suggesting that orthogonality is a strong form of linear independence. An *orthogonal basis* is one consisting of mutually orthogonal vectors.

A.4   Norms and normalization

A *norm* is a function that measures the length of vectors. In a finite-dimensional vector space, we typically use the norm $\| u \| := \langle u \mid u \rangle^{1/2}$. This norm is called the 2-norm, and it is just one of a whole family of *p*-norms that are defined as follows:

$$\| u \|_p := \left(\sum_i |u_i|^p \right)^{1/p}$$

In the limit as *p* goes to infinity, we get what is known as the *max-norm*:

$$\| u \|_\infty := \max_i |u_i|$$

If we are working with a function space (such as the functions that are continuous on [0, 1]), we ordinarily use one of the L^p norms, defined as

$$\| u \|_p := \left(\int_0^1 |u(x)|^p \, dx \right)^{1/p}$$

The max-norm for functions on the interval [0, 1] is given by

$$\| u \|_\infty := \max_{x \in [0,1]} |u(x)|$$

The most frequently used norm for functions is the L^2 norm, which can also be written as $\| u \|_2 = \langle u \mid u \rangle^{1/2}$ if we use the standard inner product.

A vector *u* with $\| u \| = 1$ is said to be *normalized*. If we have an orthogonal basis composed of vectors that are normalized in the L^2 norm, the basis is called *orthonormal*. Stated concisely, a basis $u_1, u_2, \ldots$ is orthonormal if

$$\langle u_i \mid u_j \rangle = \delta_{i,j}$$

where $\delta_{i,j}$ is called the Kronecker delta and is defined to be 1 if $i = j$ and 0 otherwise.

Example: The vectors $e_1 = [1\ 0\ 0]^T$, $e_2 = [0\ 1\ 0]^T$, $e_3 = [0\ 0\ 1]^T$ form an orthonormal basis for the inner product space $\mathbb{R}^3$ endowed with the dot product of equation (A.1). ∎

A.5 Eigenvectors and eigenvalues

A column vector v_i is said to be a *right eigenvector* of a square matrix M with associated *eigenvalue* λ_i if

$$M v_i = \lambda_i v_i \tag{A.2}$$

By convention, the eigenvectors and eigenvalues are typically ordered so that $\lambda_1 \le \lambda_2 \le \cdots \le \lambda_n$.

By juxtaposing the column vectors v_i into a single square matrix V and by placing the eigenvalues λ_i along the diagonal of a matrix $\Lambda := \mathrm{diag}(\lambda_i)$, we can rewrite equation (A.2) in matrix form:

$$M V = \Lambda V \tag{A.3}$$

A *nondefective matrix M* has n linearly independent right eigenvectors $v_1, \ldots, v_n$. Thus, the matrix of right eigenvectors V for any nondefective matrix can be inverted. Multiplying both sides of equation (A.3) by V^{-1} on both the left and the right yields

$$V^{-1}M V V^{-1} = V^{-1}\Lambda V V^{-1}$$
$$V^{-1}M = \Lambda V^{-1}$$

This equation defines the matrix of *left eigenvectors* $U := V^{-1}$, which contains a juxtaposed sequence of row vectors u_i that satisfy

$$u_i M = \lambda_i u_i \tag{A.4}$$

Because the right eigenvectors form a basis, any column vector w of height n can be written as a linear combination of the form

$$w = \sum_j a_j v_j$$

The coefficients $a_1, \ldots, a_n$ can be determined by using the left eigenvectors. In particular, since by construction $\boldsymbol{u}_i \cdot \boldsymbol{v}_j = \delta_{i,j}$, each coefficient a_i can be obtained by multiplying both sides by $\boldsymbol{u}_i$:

$$\boldsymbol{u}_i \, \boldsymbol{w} = \sum_j a_j \, \boldsymbol{u}_i \, \boldsymbol{v}_j$$

$$= \sum_j a_j \, \delta_{i,j}$$

$$= a_i$$

B-SPLINE WAVELET MATRICES

![B](decorative letter B)

*1. Haar wavelets — 2. Endpoint-interpolating linear B-spline wavelets —
3. Endpoint-interpolating quadratic B-spline wavelets — 4. Endpoint-
interpolating cubic B-spline wavelets*

This appendix presents the matrices required to make use of endpoint-interpolating B-spline wavelets of low degree. The Matlab code used to generate these matrices appears in Appendix C. These concrete examples should serve to elucidate the ideas presented in Section 7.3.2. In order to emphasize the sparse structure of the matrices, zeros have been omitted. Diagonal dots indicate that the previous column is to be repeated the appropriate number of times, shifted down by two rows for each column. The P matrices have entries relating unnormalized B-spline scaling functions, while the Q matrices have entries defining normalized, minimally supported wavelets. Columns of the Q matrices that are not represented exactly with integers are given to six decimal places.

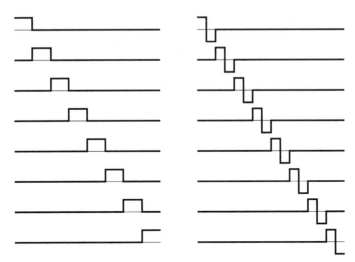

FIGURE B.1   The piecewise-constant B-spline scaling functions and wavelets for $j = 3$.

B.1   Haar wavelets

The B-spline wavelet basis of degree 0 is simply the Haar basis described in Chapter 2. Some examples of the box scaling functions and Haar wavelets are depicted in Figure B.1. The synthesis matrices P^j and Q^j are given below.

$$P^j = \begin{bmatrix} 1 & & & & \\ 1 & & & & \\ & 1 & & & \\ & 1 & \cdot & & \\ & & & \ddots & \\ & & & \cdot & 1 \\ & & & & 1 \end{bmatrix} \qquad Q^j = \sqrt{\frac{2^j}{2}} \begin{bmatrix} 1 & & & & \\ -1 & & & & \\ & 1 & & & \\ & -1 & \cdot & & \\ & & & \ddots & \\ & & & \cdot & 1 \\ & & & & -1 \end{bmatrix}$$

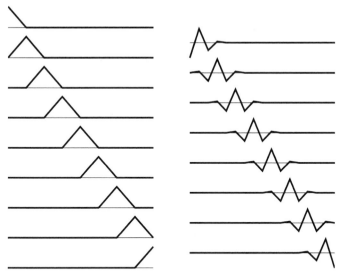

FIGURE B.2   The linear B-spline scaling functions and wavelets for $j = 3$.

B.2   Endpoint-interpolating linear B-spline wavelets

Figure B.2 shows a few typical scaling functions and wavelets for linear B-splines. The synthesis matrices $\boldsymbol{P}^j$ and $\boldsymbol{Q}^j$ for endpoint-interpolating linear B-spline wavelets are given below.

$$\boldsymbol{P}^1 = \frac{1}{2}\begin{bmatrix} 2 & \\ 1 & 1 \\ & 2 \end{bmatrix} \qquad \boldsymbol{Q}^1 = \sqrt{3}\begin{bmatrix} -1 \\ 1 \\ -1 \end{bmatrix}$$

$$\boldsymbol{P}^2 = \frac{1}{2}\begin{bmatrix} 2 & & \\ 1 & 1 & \\ & 2 & \\ & 1 & 1 \\ & & 2 \end{bmatrix} \qquad \boldsymbol{Q}^2 = \sqrt{\frac{3}{64}}\begin{bmatrix} -12 & \\ 11 & 1 \\ -6 & -6 \\ 1 & 11 \\ & -12 \end{bmatrix}$$

$$P^{j \geq 3} = \frac{1}{2} \begin{bmatrix} 2 & & & & & & & & \\ 1 & 1 & & & & & & & \\ & 2 & & & & & & & \\ & 1 & 1 & & & & & & \\ & & 2 & & & & & & \\ & & 1 & \cdot & & & & & \\ & & & & \cdot & & & & \\ & & & & & \cdot & 1 & & \\ & & & & & & 2 & & \\ & & & & & & 1 & 1 & \\ & & & & & & & 2 & \end{bmatrix} \qquad Q^{j \geq 3} = \sqrt{\frac{2^j}{72}} \begin{bmatrix} -11.022704 & & & \\ 10.104145 & 1 & & \\ -5.511352 & -6 & & \\ 0.918559 & 10 & 1 & \\ & -6 & -6 & \\ & 1 & 10 & \\ & & -6 & \cdot \\ & & 1 & \cdot & 1 \\ & & & \cdot & -6 \\ & & & 10 & 0.918559 \\ & & & -6 & -5.511352 \\ & & & 1 & 10.104145 \\ & & & & -11.022704 \end{bmatrix}$$

B.3 Endpoint-interpolating quadratic B-spline wavelets

Figure B.3 shows some of the quadratic B-spline scaling functions and wavelets. The synthesis matrices P^j and Q^j for the quadratic case are given below.

$$P^1 = \frac{1}{2} \begin{bmatrix} 2 & \\ 1 & 1 \\ & 1 & 1 \\ & & 2 \end{bmatrix} \qquad Q^1 = \sqrt{\frac{5}{4}} \begin{bmatrix} -2 \\ 3 \\ -3 \\ 2 \end{bmatrix}$$

$$P^2 = \frac{1}{4} \begin{bmatrix} 4 & & \\ 2 & 2 & \\ & 3 & 1 \\ & 1 & 3 \\ & & 2 & 2 \\ & & & 4 \end{bmatrix} \qquad Q^2 = \sqrt{\frac{3}{4936}} \begin{bmatrix} -144 & \\ 177 & 21 \\ -109 & -53 \\ 53 & 109 \\ -21 & -177 \\ & 144 \end{bmatrix}$$

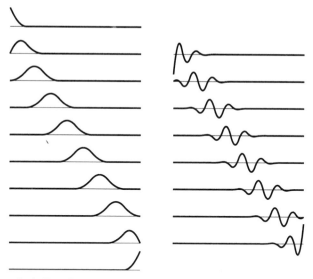

FIGURE B.3   The quadratic B-spline scaling functions and wavelets for $j = 3$.

$$P^{j \geq 3} = \frac{1}{4} \begin{bmatrix} 4 & & & & & & & & \\ 2 & 2 & & & & & & & \\ & 3 & 1 & & & & & & \\ & 1 & 3 & & & & & & \\ & & 3 & 1 & & & & & \\ & & 1 & 3 & & & & & \\ & & & 3 & \cdot & & & & \\ & & & 1 & \cdot & 1 & & & \\ & & & & \cdot & 3 & & & \\ & & & & & 3 & 1 & & \\ & & & & & 1 & 3 & & \\ & & & & & & 2 & 2 & \\ & & & & & & & 4 & \end{bmatrix}$$

$$Q^3 = \sqrt{\frac{1}{713568}} \begin{bmatrix} -4283.828550 & & & \\ 5208.746077 & 780 & & \\ -3099.909150 & -1949 & -11 & \\ 1300.002166 & 3481 & 319 & \\ -253.384964 & -3362 & -1618 & -8.737413 \\ 8.737413 & 1618 & 3362 & 253.384964 \\ & -319 & -3481 & -1300.002166 \\ & 11 & 1949 & 3099.909150 \\ & & -780 & -5208.746077 \\ & & & 4283.828550 \end{bmatrix}$$

$$
Q^{j \geq 4} = \sqrt{\frac{3 \cdot 2^j}{136088}}
\begin{bmatrix}
-381.872771 & & & & & & & & & \\
464.322574 & 69.531439 & & & & & & & & \\
-276.334798 & -173.739454 & -1 & & & & & & & \\
115.885924 & 310.306330 & 29 & & & & & & & \\
-22.587463 & -299.698329 & -147 & -1 & & & & & & \\
0.778878 & 144.233164 & 303 & 29 & & & & & & \\
& -28.436576 & -303 & -147 & & & & & & \\
& 0.980572 & 147 & 303 & & -1 & & & & \\
& & -29 & -303 & \cdot & 29 & & & & \\
& & 1 & 147 & \cdot & -147 & -0.980572 & & & \\
& & & -29 & \cdot & 303 & 28.436576 & & & \\
& & & 1 & & -303 & -144.233164 & -0.77878 & & \\
& & & & & 147 & 299.698329 & 22.587463 & & \\
& & & & & -29 & -310.306330 & -115.885924 & & \\
& & & & & 1 & 173.739454 & 276.334798 & & \\
& & & & & & -69.531439 & -464.322574 & & \\
& & & & & & & 381.872771 &
\end{bmatrix}
$$

B.4   Endpoint-interpolating cubic B-spline wavelets

Some examples of cubic B-spline scaling functions and wavelets are shown in Figure B.4. The synthesis matrices P^j and Q^j for endpoint-interpolating cubic B-spline wavelets are given below.

$$
P^1 = \frac{1}{2}
\begin{bmatrix}
2 & & \\
1 & 1 & \\
& 1 & 1 \\
& & 1 & 1 \\
& & & 2
\end{bmatrix}
\qquad
Q^1 = \sqrt{7}
\begin{bmatrix}
1 \\
-2 \\
3 \\
-2 \\
1
\end{bmatrix}
$$

$$
P^2 = \frac{1}{16}
\begin{bmatrix}
16 & & \\
8 & 8 & \\
& 12 & 4 \\
& 3 & 10 & 3 \\
& & 4 & 12 \\
& & & 8 & 8 \\
& & & & 16
\end{bmatrix}
\qquad
Q^2 = \sqrt{\frac{315}{31196288}}
\begin{bmatrix}
1368 & \\
-2064 & -240 \\
1793 & 691 \\
-1053 & -1053 \\
691 & 1793 \\
-240 & -2064 \\
& 1368
\end{bmatrix}
$$

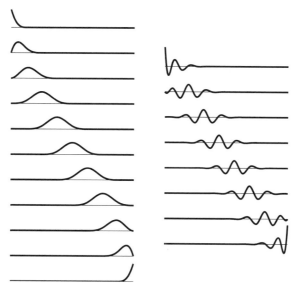

FIGURE B.4   The cubic B-spline scaling functions and wavelets for $j = 3$.

$$
P^{j \geq 3} = \frac{1}{16}
\begin{bmatrix}
16 & & & & & & & \\
8 & 8 & & & & & & \\
 & 12 & 4 & & & & & \\
 & 3 & 11 & 2 & & & & \\
 & & 8 & 8 & & & & \\
 & & 2 & 12 & 2 & & & \\
 & & & 8 & 8 & & & \\
 & & & 2 & 12 & & & \\
 & & & & 8 & \cdot & & \\
 & & & & 2 & \cdot & 2 & \\
 & & & & & \cdot & 8 & \\
 & & & & & & 12 & 2 \\
 & & & & & & 8 & 8 \\
 & & & & & & 2 & 11 & 3 \\
 & & & & & & & 4 & 12 \\
 & & & & & & & & 8 & 8 \\
 & & & & & & & & & 16
\end{bmatrix}
\qquad
Q^3 =
\begin{bmatrix}
6.311454 & & & \\
-9.189342 & -1.543996 & & \\
7.334627 & 4.226722 & 0.087556 & \\
-3.514553 & -5.585477 & -0.473604 & -0.000155 \\
1.271268 & 6.059557 & 1.903267 & 0.019190 \\
-0.259914 & -4.367454 & -4.367454 & -0.259914 \\
0.019190 & 1.903267 & 6.059557 & 1.271268 \\
-0.000155 & -0.473604 & -5.585477 & -3.514553 \\
 & 0.087556 & 4.226722 & 7.334627 \\
 & & -1.543996 & -9.189342 \\
 & & & 6.311454
\end{bmatrix}
$$

$$\varrho^{j\geq 4} = \sqrt{\dfrac{5\cdot 2^{j}}{675221664}}$$

$$
\begin{bmatrix}
25931.200710 \\
-37755.271723 & -6369.305453 \\
30135.003012 & 17429.266054 & 385.797044 \\
-14439.869635 & -23004.252368 & -2086.545605 & -1 \\
5223.125428 & 24848.487871 & 8349.373420 & 124 \\
-1067.879425 & -17678.884301 & -18743.473059 & -1677 & -1 \\
78.842887 & 7394.685374 & 24291.795239 & 7904 & 124 \\
-0.635830 & -1561.868558 & -18420.997597 & -18482 & -1677 \\
 & 115.466347 & 7866.732009 & 24264 & 7904 \\
 & -0.931180 & -1668.615872 & -18482 & -18482 & -1 \\
 & & 123.378671 & 7904 & 24264 & 124 \\
 & & -0.994989 & -1677 & -18482 & -1677 & -0.994989 \\
 & & & 124 & 7904 & 7904 & 123.378671 \\
 & & & -1 & -1677 & -18482 & -1668.615872 & -0.931180 \\
 & & & & 124 & 24264 & 7866.732009 & 115.466347 \\
 & & & & -1 & -18482 & -18420.997597 & -1561.868558 & -0.635830 \\
 & & & & & 7904 & 24291.795239 & 7394.685374 & 78.842887 \\
 & & & & & -1677 & -18743.473059 & -17678.884301 & -1067.879425 \\
 & & & & & 124 & 8349.373420 & 24848.487871 & 5223.125428 \\
 & & & & & -1 & -2086.545605 & -23004.252368 & -14439.869635 \\
 & & & & & & 385.797044 & 17429.266054 & 30135.003012 \\
 & & & & & & & -6369.305453 & -37755.271723 \\
 & & & & & & & & 25931.200710
\end{bmatrix}
$$

MATLAB CODE FOR
B-SPLINE WAVELETS

This appendix presents algorithms for generating the synthesis matrices P^j and Q^j for B-spline wavelets of arbitrary degree. The implementation here is for Matlab, but the same results could be obtained with any matrix library or package.

For a given B-spline degree d and hierarchy level j, the matrices P^j and Q^j are created by the following commands:

```
P = FindP(d, j);
Q = FindQ(d, j);
```

The scaling function synthesis matrix P^j is determined using *knot insertion* [35]. The B-spline basis functions described in Section 7.3 are defined over a knot sequence consisting of multiple knots at the endpoints and equally spaced simple knots on the interior. By inserting new knots midway between each pair of interior knots, we can determine how the scaling functions at level $j-1$ can be written in terms of those at level j.

```
function P = FindP(d, j)
% P = FindP(d, j) returns the P matrix for B-spline scaling functions of
% degree d, level j.
d = fix(d);
if d < 0,
```

```matlab
    error('FindP: Must have d >= 0.');
end;
j = fix(j);
if j < 1
    error('FindP: Must have j >= 1.');
end;
if d == 0
  P = [1; 1];
  for i = 2:j
    P = [P zeros(size(P)); zeros(size(P)) P];
  end;
else
  u = Knots(d, j - 1);
  g = Greville(d, u);
  P = eye(2^(j - 1) + d);
  for k = 0:2^(j - 1)-1
    [u, g, P] = InsertKnot(d, u, g, P, (2*k+1)/2^j);
  end;
end;
return;


function x = Knots(d, j)
% x = Knots(d, j) returns a vector of knot values for B-spline scaling
% functions of degree d, level j.
x = [zeros(1, d-1) [0:2^j-1]/2^j ones(1,d)];
return;


function x = Greville(d, u)
% x = Greville(d, u) returns the vector of Greville abscissa values
% corresponding to degree d and knot vector u.
l = length(u);
x = u(1:l-d+1);
for k = 2:d
  x = x + u(k:l-d+k);
end;
x = x / d;
return;
```

```
function [uret, gret, pret] = InsertKnot(d, u, g, p, unew)
% [uret, gret, pret] = InsertKnot(d, u, g, p, unew) inserts a new knot
% at unew for B-spline scaling functions of degree d, thereby modifying
% knot vector u, Greville abscissas g, and synthesis matrix p.
uret = sort([u unew]);
gret = Greville(d, uret);
pret = PolyEval(g, p, gret);
return;


function pret = PolyEval(g, p, gnew)
% pret = PolyEval(g, p, gnew) returns the values of a control polygon
% defined by abscissas g and ordinates p, evaluated at gnew.
[m, n] = size(p);
if length(g) ~= m
  error('PolyEval: Length of g and rows of p must be the same.');
end;
for i = 1:length(gnew)
  row = max(find(g <= gnew(i)));
  if row == m
    pret(i,:) = p(m,:);
  else
    frac = (g(row+1) - gnew(i))/(g(row+1) - g(row));
    pret(i,:) = frac*p(row,:) + (1-frac)*p(row+1,:);
  end;
end;
return;
```

The matrix of inner products of B-spline scaling functions with one another can be computed by writing each basis function as a linear combination of Bernstein polynomials; inner products of Bernstein polynomials are relatively simple. The weights of the Bernstein polynomials are found by repeatedly inserting knots at interior knots until they acquire multiplicity $d + 1$.

```
function I = Inner(d, j)
% I = Inner(d, j) returns the inner product matrix for B-spline scaling
% functions of degree d at level j.
I0 = BernsteinInner(d);
n = 2^j + d;
I = zeros(n);
```

```
w = BernsteinWeights(d, j);
for k = 1:n
  w1 = reshape(w(:,k), d+1, 2^j);
  for l = k:n
    w2 = reshape(w(:,l), d+1, 2^j);
    I(k,l) = trace(w1'*I0*w2);
    I(l,k) = I(k,l);
  end;
end;
I = I / 2^j;
return;


function I = BernsteinInner(d)
% I = BernsteinInner(d) returns the matrix of inner products of
% Bernstein polynomials of degree d.
i = ones(d+1, 1)*[0:d];
j = i';
I = Choose(d, i).*Choose(d, j)./(Choose(2*d, i+j)*(2*d + 1));
return;


function c = Choose(n, r)
% c = Choose(n, r) returns (n choose r) = n! / (r! (n-r)!).
c = Factorial(n)./(Factorial(r).*Factorial(n-r));
return;


function f = Factorial(m)
% f = Factorial(m) returns the matrix of factorials of entries of m.
[r,c] = size(m);
f = zeros(r, c);
for i = 1:r
  for j = 1:c
    f(i,j) = prod(2:m(i,j));
  end;
end;
return;
```

```
function w = BernsteinWeights(d, j)
% w = BernsteinWeights(d, j) returns a matrix of B-spline scaling
% function weights for Bernstein polynomials of degree d, level j.
w = eye(2^j + d);
if d == 0
  return;
end;
u = Knots(d, j);
g = Greville(d, u);
for i = 1:2^j - 1
  for r = 1:d
    [u, g, w] = InsertKnot(d, u, g, w, i/2^j);
  end;
end;
return;
```

The final function determines Q^j one column at a time, using knowledge of the small supports of B-spline wavelets.

```
function Q = FindQ(d, j, normalization)
% Q = FindQ(d, j, normalization) returns the Q matrix for B-spline
% scaling functions of degree d, level j. If normalization is 'min' (or
% is not specified) then the smallest entry in each column is made 1. If
% normalization is 'max' then the largest entry in each column is made
% 1. If normalization is 'L2' then the L^2 norm of each wavelet is
% made 1.
if nargin < 3
  normalization = 'min';
elseif ~strcmp(normalization, 'min') & ~strcmp(normalization, 'max') ...
      & ~strcmp(normalization, 'L2')
  error('FindQ: normalization must be ''min'', ''max'', or ''L2''.');
end;
P = FindP(d, j);
I = Inner(d, j);
M = P'*I;
[m1, m2] = size(M);
n = m2 - rank(M);
Q = zeros(m2, n);
```

```
found = 0;
start_col = 0;
while (found < n/2) & (start_col < m2)
  start_col = start_col + 1 + (found > d);
  width = 0;
  rank_def = 0;
  while ~rank_def & (width < m2 - start_col + 1)
    width = width + 1;
    submatrix = M(:,start_col:start_col+width-1);
    rank_def = width - rank(submatrix);
end;
if rank_def
% find null space of submatrix (should be just one column)
q_col = null(submatrix);
if strcmp(normalization, 'min')
  % normalize column so smallest nonzero entry has magnitude 1
  q_col = q_col/min(abs(q_col + 1e38*(abs(q_col) < 1e-10)));
elseif strcmp(normalization, 'max')
  % normalize column so largest entry has magnitude 1
  q_col = q_col/max(abs(q_col));
end;
    % change sign to give consistent orientation
    q_col = q_col*(-1)^(start_col + floor((d+1)/2) + (q_col(1,1) > 0));
    % put column into left half of Q
    found = found + 1;
    Q(start_col:start_col + width-1,found) = q_col;
    % use symmetry to put column into right half of Q in reverse order
    % and negated if degree is even
    Q(:,n-found+1) = flipud(Q(:,found))*(-1)^(d+1);
  end;
end;
if strcmp(normalization, 'L2')
  % normalize matrix so each column has L^2 norm of 1
  ip = Q'*I*Q;
  Q = Q*diag(1./sqrt(diag(ip)));
end;
return;
```

BIBLIOGRAPHY

[1] Michael J. Banks and Elaine Cohen. Realtime spline curves from interactively sketched data. *Computer Graphics,* 24(2):99–107, March 1990.

[2] R. H. Bartels and J. C. Beatty. A technique for the direct manipulation of spline curves. In *Proceedings of the 1989 Graphics Interface Conference,* pages 33–39. Canadian Information Processing Society, London, Ontario, Canada, 1989.

[3] R. Bartels, J. Beatty, and B. Barsky. *An Introduction to Splines for Use in Computer Graphics and Geometric Modeling.* Morgan Kaufmann, San Francisco, 1987.

[4] Deborah Berman, Jason Bartell, and David Salesin. Multiresolution painting and compositing. In *Proceedings of SIGGRAPH '94,* pages 85–90. ACM, New York, 1994.

[5] G. Beylkin, R. Coifman, and V. Rokhlin. Fast wavelet transforms and numerical algorithms I. *Communications on Pure and Applied Mathematics,* 44(2):141–183, March 1991.

[6] Peter J. Burt and Edward H. Adelson. The Laplacian pyramid as a compact image code. *IEEE Transactions on Communications,* 31(4):532–540, April 1983.

[7] E. Catmull and J. Clark. Recursively generated B-spline surfaces on arbitrary topological meshes. *Computer-Aided Design,* 10(6):350–355, November 1978.

[8] Alfred S. Cavaretta, Wolfgang Dahmen, and Charles A. Micchelli. *Stationary Subdivision.* American Mathematical Society, Providence, RI, 1991.

[9]   George Celniker and Dave Gossard. Deformable curve and surface finite elements for free-form shape design. In *Proceedings of SIGGRAPH '91,* pages 257–265. ACM, New York, 1991.

[10] Andrew Certain, Jovan Popović, Tony DeRose, Tom Duchamp, David Salesin, and Werner Stuetzle. Interactive multiresolution surface viewing. In *Proceedings of SIGGRAPH '96.* ACM, New York, 1996 (to appear).

[11] George M. Chaikin. An algorithm for high speed curve generation. *Computer Graphics and Image Processing,* 3(4):346–349, December 1974.

[12] Per H. Christensen. Hierarchical techniques for glossy global illumination. Ph.D. thesis, University of Washington, 1995.

[13] Per H. Christensen, Dani Lischinski, Eric J. Stollnitz, and David H. Salesin. Clustering for glossy global illumination. *ACM Transactions on Graphics,* 1996 (to appear).

[14] Per H. Christensen, Eric J. Stollnitz, David H. Salesin, and Tony D. DeRose. Wavelet radiance. In G. Sakas, P. Shirley, and S. Müller, editors, *Photorealistic Rendering Techniques,* pages 295–309. Springer-Verlag, Berlin, 1995.

[15] Per H. Christensen, Eric J. Stollnitz, David H. Salesin, and Tony D. DeRose. Global illumination of glossy environments using wavelets and importance. *ACM Transactions on Graphics,* 15(1):37–71, January 1996.

[16] H. N. Christiansen and T. W. Sederberg. Conversion of complex contour line definitions into polygonal element mosaics. In *Proceedings of SIGGRAPH '78,* pages 187–192. ACM, New York, 1978.

[17] Charles K. Chui. *An Introduction to Wavelets.* Academic Press, Boston, 1992.

[18] Charles K. Chui and Ewald Quak. Wavelets on a bounded interval. In D. Braess and L. L. Schumaker, editors, *Numerical Methods in Approximation Theory,* volume 9, pages 53–75. Birkhauser Verlag, Basel, 1992.

[19] Hsiung Chuan Chun and C. C. J. Kuo. Contour metamorphosis using the wavelet descriptor. In *Image and Video Processing II,* volume 2182 of *Proceedings of the SPIE,* pages 288–299. SPIE, Bellingham, WA, 1994.

[20] A. Cohen, I. Daubechies, and J. C. Feauveau. Biorthogonal bases of compactly supported wavelets. *Communications on Pure and Applied Mathematics,* 45(5):485–500, June 1992.

[21] Michael F. Cohen. Interactive spacetime control for animation. In *Proceedings of SIGGRAPH '92,* pages 293–302. ACM, New York, 1992.

[22] Michael F. Cohen and John R. Wallace. *Radiosity and Realistic Image Synthesis.* Academic Press Professional, Cambridge, MA, 1993.

[23] A. Croisier, D. Esteban, and C. Galand. Perfect channel splitting by use of inter-polation/decimation/tree decomposition techniques. In *International Conference on Information Sciences and Systems,* pages 443–446. Hemisphere, Washington, DC, 1977.

[24] Ingrid Daubechies. Orthonormal bases of compactly supported wavelets. *Communications on Pure and Applied Mathematics,* 41(7):909–996, October 1988.

[25] Ingrid Daubechies. *Ten Lectures on Wavelets.* SIAM, Philadelphia, 1992.

[26] Tony D. DeRose, Michael Lounsbery, and Leena-Maija Reissell. Curves and surfaces. In Alain Fournier, editor, *SIGGRAPH '95 Course Notes 26: Wavelets and Their Applications in Computer Graphics,* pages 123–154. ACM, New York, 1995.

[27] R. DeVore, B. Jawerth, and B. Lucier. Image compression through wavelet transform coding. *IEEE Transactions on Information Theory,* 38(2):719–746, March 1992.

[28] R. A. DeVore, B. Jawerth, and B. J. Lucier. Surface compression. *Computer-Aided Geometric Design,* 9(3):219–239, August 1992.

[29] D. W. H. Doo. A recursive subdivision algorithm for fitting quadratic surfaces to irregular polyhedrons. Ph.D. thesis, Brunel University, 1978.

[30] D. Doo and M. Sabin. Behaviour of recursive division surfaces near extraordinary points. *Computer-Aided Design,* 10(6):356–360, September 1978.

[31] Nira Dyn and David Levin. The subdivision experience. In P.-J. Laurent, A. Le Méhauté, and L. L. Schumaker, editors, *Wavelets, Images, and Surface Fitting,* pages 229–244. A. K. Peters, Wellesley, MA, 1994.

[32] Nira Dyn, David Levin, and John Gregory. A four-point interpolatory subdivision scheme for curve design. *Computer-Aided Geometric Design,* 4(4):257–268, December 1987.

[33] Nira Dyn, David Levin, and John Gregory. A butterfly subdivision scheme for surface interpolation with tension control. *ACM Transactions on Graphics,* 9(2):160–169, April 1990.

[34] Matthias Eck, Tony DeRose, Tom Duchamp, Hugues Hoppe, Michael Lounsbery, and Werner Stuetzle. Multiresolution analysis of arbitrary meshes. In *Proceedings of SIGGRAPH '95,* pages 173–182. ACM, New York, 1995.

[35] Gerald Farin. *Curves and Surfaces for Computer Aided Geometric Design.* Academic Press, Boston, third edition, 1993.

[36] Adam Finkelstein, Charles E. Jacobs, and David H. Salesin. Multiresolution video. In *Proceedings of SIGGRAPH '96.* ACM, New York, 1996 (to appear).

[37] Adam Finkelstein and David H. Salesin. Multiresolution curves. In *Proceedings of SIGGRAPH '94,* pages 261–268. ACM, New York, 1994.

[38] James D. Foley, Andries van Dam, Steven K. Feiner, and John F. Hughes. *Computer Graphics: Principles and Practice.* Addison-Wesley, Reading, MA, second edition, 1990.

[39] David R. Forsey and Richard H. Bartels. Hierarchical B-spline refinement. In *Proceedings of SIGGRAPH '88,* pages 205–212. ACM, New York, 1988.

[40] David R. Forsey and Richard H. Bartels. Surface fitting with hierarchical splines. *ACM Transactions on Graphics,* 14(2):134–161, April 1995.

[41] D. Forsey and R. Bartels. Tensor products and hierarchical fitting. In *Curves and Surfaces in Computer Vision and Graphics II,* volume 1610 of *Proceedings of the SPIE,* pages 88–96. SPIE, Bellingham, WA, 1991.

[42] Barry Fowler. Geometric manipulation of tensor product surfaces. In *Proceedings of the 1992 Symposium on Interactive 3D Graphics,* pages 101–108. ACM, New York, 1992.

[43] H. Fuchs, Z. M. Kedem, and S. P. Uselton. Optimal surface reconstruction from planar contours. *Communications of the ACM,* 20(10):693–702, October 1977.

[44] Dennis Gabor. Theory of communication. *Journal of the Institute of Electrical Engineers,* 93(22):429–457, 1946.

[45] S. Ganapathy and T. G. Dennehy. A new general triangulation method for planar contours. In *Proceedings of SIGGRAPH '82,* pages 69–75. ACM, New York, 1982.

[46] Andrew S. Glassner. *Principles of Digital Image Synthesis,* volume II. Morgan Kaufmann, San Francisco, 1995.

[47] Gene H. Golub and Charles F. Van Loan. *Matrix Computations.* The Johns Hopkins University Press, Baltimore, second edition, 1989.

[48] Yihong Gong, Hongjiang Zhang, H. C. Chuan, and M. Sakauchi. An image database system with content capturing and fast image indexing abilities. In *Proceedings of the International Conference on Multimedia Computing and Systems,* pages 121–130. IEEE Computer Society Press, Los Alamitos, CA, 1994.

[49] Steven J. Gortler and Michael F. Cohen. Hierarchical and variational geometric modeling with wavelets. In *Proceedings of the 1995 Symposium on Interactive 3D Graphics,* pages 35–42. ACM, New York, 1995.

[50] Steven J. Gortler, Peter Schröder, Michael F. Cohen, and Pat Hanrahan. Wavelet radiosity. In *Proceedings of SIGGRAPH '93,* pages 221–230. ACM, New York, 1993.

[51] Byron S. Gottfried and Joel Weisman. *Introduction to Optimization Theory.* Prentice Hall, Englewood Cliffs, NJ, 1973.

[52] A. Grossman and J. Morlet. Decomposition of functions into wavelets of constant shape, and related transforms. In L. Streit, editor, *Mathematics and Physics: Lectures on Recent Results.* World Scientific, Singapore, 1985.

[53] Alfred Haar. Zur Theorie der orthogonalen Funktionen-Systeme. *Mathematische Annalen,* 69:331–371, 1910.

[54] Mark Halstead, Michael Kass, and Tony DeRose. Efficient, fair interpolation using Catmull-Clark surfaces. In *Proceedings of SIGGRAPH '93,* pages 35–44. ACM, New York, 1993.

[55] Pat Hanrahan, David Salzman, and Larry Aupperle. A rapid hierarchical radiosity algorithm. In *Proceedings of SIGGRAPH '91,* pages 197–206. ACM, New York, 1991.

[56] Taosong He, Sidney Wang, and Arie Kaufman. Wavelet-based volume morphing. In *Proceedings of Visualization '94,* pages 85–92. IEEE Computer Society Press, Los Alamitos, CA, 1994.

[57] Donald Hearn and M. Pauline Baker. *Computer Graphics.* Addison-Wesley, Reading, MA, 1994.

[58] David J. Heeger and James R. Bergen. Pyramid-based texture analysis/synthesis. In *Proceedings of SIGGRAPH '95,* pages 229–238. ACM, New York, 1995.

[59] K. Hirata and T. Kato. Query by visual example—content based image retrieval. In A. Pirotte, C. Delobel, and G. Gottlob, editors, *Advances in Database Technology (EDBT '92),* pages 56–71. Springer-Verlag, Berlin, 1992.

[60] H. Hoppe, T. DeRose, T. Duchamp, M. Halstead, H. Jin, J. McDonald, J. Schweitzer, and W. Stuetzle. Piecewise smooth surface reconstruction. In *Proceedings of SIGGRAPH '94,* pages 295–302. ACM, New York, 1994.

[61] Josef Hoschek and Dieter Lasser. *Fundamentals of Computer Aided Geometric Design.* A. K. Peters, Wellesley, MA, third edition, 1993.

[62] William M. Hsu, John F. Hughes, and Henry Kaufman. Direct manipulation of free-form deformations. In *Proceedings of SIGGRAPH '92,* pages 177–184. ACM, New York, 1992.

[63] Charles E. Jacobs, Adam Finkelstein, and David H. Salesin. Fast multiresolution image querying. In *Proceedings of SIGGRAPH '95,* pages 277–286. ACM, New York, 1995.

[64] Rong-Qing Jia and Charles A. Micchelli. Using the refinement equations for the construction of pre-wavelets II: Powers of two. In P.-J. Laurent, A. Le Méhauté, and L. L. Schumaker, editors, *Curves and Surfaces,* pages 209–246. Academic Press, Boston, 1991.

[65] Atreyi Kankanhalli, Hong Jiang Zhang, and Chien Yong Low. Using texture for image retrieval. In *International Conference on Automation, Robotics and Computer Vision.* Nanyang Technological University, Singapore, 1994.

[66] T. Kato, T. Kurita, N. Otsu, and K. Hirata. A sketch retrieval method for full color image database—query by visual example. In *Proceedings of the 11th IAPR International Conference on Pattern Recognition,* pages 530–533. IEEE Computer Society Press, Los Alamitos, CA, 1992.

[67] E. Keppel. Approximating complex surfaces by triangulation of contour lines. *IBM Journal of Research and Development,* 19(1):2–11, January 1975.

[68] A. Kitamoto, C. Zhou, and M. Takagi. Similarity retrieval of NOAA satellite imagery by graph matching. In *Storage and Retrieval for Image and Video Databases,* volume 1908 of *Proceedings of the SPIE,* pages 60–73. SPIE, Bellingham, WA, 1993.

[69] J. Lane and R. Riesenfeld. A theoretical development for the computer generation and display of piecewise polynomial surfaces. *IEEE Transactions on Pattern Analysis and Machine Intelligence,* 2(1):35–46, January 1980.

[70] A. S. Lewis and G. Knowles. Video compression using 3D wavelet transforms. *Electronics Letters,* 26(6):396–398, 15 March 1990.

[71] Dani Lischinski, Filippo Tampieri, and Donald P. Greenberg. Combining hierarchical radiosity and discontinuity meshing. In *Proceedings of SIGGRAPH '93,* pages 199–208. ACM, New York, 1993.

[72] Zicheng Liu, Steven J. Gortler, and Michael F. Cohen. Hierarchical spacetime control. In *Proceedings of SIGGRAPH '94,* pages 35–42. ACM, New York, 1994.

[73] Charles T. Loop. Smooth subdivision surfaces based on triangles. Master's thesis, Department of Mathematics, University of Utah, 1987.

[74] J. Michael Lounsbery. Multiresolution analysis for surfaces of arbitrary topological type. Ph.D. thesis, University of Washington, 1994.

[75] Michael Lounsbery, Tony DeRose, and Joe Warren. Multiresolution surfaces of arbitrary topological type. *ACM Transactions on Graphics,* 1996 (to appear).

[76] T. Lyche and K. Mørken. Knot removal for parametric B-spline curves and surfaces. *Computer-Aided Geometric Design,* 4(3):217–230, November 1987.

[77] Stephane Mallat. Multiresolution representation and wavelets. Ph.D. thesis, University of Pennsylvania, 1988.

[78] Stephane Mallat. A theory for multiresolution signal decomposition: The wavelet representation. *IEEE Transactions on Pattern Analysis and Machine Intelligence,* 11(7):674–693, July 1989.

[79] Stephane Mallat and Sifen Zhong. Wavelet transform maxima and multiscale edges. In M. B. Ruskai et al., editors, *Wavelets and Their Applications,* pages 67–104. Jones and Bartlett, Boston, 1992.

[80] Yves Meyer. Ondelettes et fonctions splines. Technical report, Séminaire EDP, École Polytechnique, Paris, 1986.

[81] Yves Meyer. Ondelettes sur l'intervalle. *Revista Matematica Iberoamericana,* 7(2):115–143, 1991.

[82] Yves Meyer. *Wavelets: Algorithms and Applications.* SIAM, Philadelphia, 1993. Translated by Robert D. Ryan.

[83] David Meyers. Multiresolution tiling. *Computer Graphics Forum,* 13(5):325–340, December 1994.

[84] David Meyers. Reconstruction of surfaces from planar contours. Ph.D. thesis, University of Washington, 1994.

[85] David Meyers, Shelley Skinner, and Kenneth Sloan. Surfaces from contours. *ACM Transactions on Graphics,* 11(3):228–258, July 1992.

[86] J. Morlet, G. Arens, E. Fourgeau, and D. Giard. Wave propagation and sampling theory. *Geophysics,* 47(2):203–236, February 1982.

[87] Philip Morrison, Phylis Morrison, and The Office of Charles and Ray Eames. *Powers of Ten: A Book About the Relative Size of Things in the Universe and the Effect of Adding Another Zero.* Scientific American, New York, 1982.

[88] Shigeru Muraki. Approximation and rendering of volume data using wavelet transforms. In *Proceedings of Visualization '92,* pages 21–28. IEEE Computer Society Press, Los Alamitos, CA, 1992.

[89] Shigeru Muraki. Volume data and wavelet transforms. *IEEE Computer Graphics and Applications,* 13(4):50–56, July 1993.

[90] W. Niblack, R. Barber, W. Equitz, M. Flickner, E. Glasman, D. Petkovic, P. Yanker, C. Faloutsos, and G. Taubin. The QBIC project: Querying images by content using color, texture, and shape. In *Storage and Retrieval for Image and Video Databases,* volume 1908 of *Proceedings of the SPIE,* pages 173–187. SPIE, Bellingham, WA, 1993.

[91]   S. Panchanathan, E. Chan, and X. Wang. Fast multiresolution motion estimation scheme for a wavelet transform video coder. In *Visual Communications and Image Processing '94,* volume 2308, part 1 of *Proceedings of the SPIE,* pages 671–681. SPIE, Bellingham, WA, 1994.

[92]   Alex P. Pentland. Equilibrium and interpolation solutions using wavelet bases. In N. M. Patrikalakis, editor, *Scientific Visualization of Physical Phenomena,* pages 507–524. Springer-Verlag, Tokyo, 1991.

[93]   Ken Perlin and David Fox. Pad: An alternative approach to the user interface. In *Proceedings of SIGGRAPH '93,* pages 57–64. ACM, New York, 1993.

[94]   Ken Perlin and Luiz Velho. Live paint: Painting with procedural multiscale textures. In *Proceedings of SIGGRAPH '95,* pages 153–160. ACM, New York, 1995.

[95]   Michael Plass and Maureen Stone. Curve-fitting with piecewise parametric cubics. In *Proceedings of SIGGRAPH '83,* pages 229–239. ACM, New York, 1983.

[96]   Thomas Porter and Tom Duff. Compositing digital images. In *Proceedings of SIGGRAPH '84,* pages 253–259. ACM, New York, 1984.

[97]   William H. Press, Brian P. Flannery, Saul A. Teukolsky, and William T. Fetterling. *Numerical Recipes.* Cambridge University Press, New York, second edition, 1992.

[98]   Ewald Quak and Norman Weyrich. Decomposition and reconstruction algorithms for spline wavelets on a bounded interval. *Applied and Computational Harmonic Analysis,* 1(3):217–231, June 1994.

[99]   J. N. Reddy. *Energy and Variational Methods in Applied Mechanics.* John Wiley & Sons, New York, 1984.

[100] Mark C. Reichert. A two-pass radiosity method driven by lights and viewer position. Master's thesis, Program of Computer Graphics, Cornell University, 1992.

[101] Ulrich Reif. A unified approach to subdivision algorithms. Technical report A-92-16, Universität Stuttgart, 1992.

[102] Norman Ricker. The form and nature of seismic waves and the structure of seismograms. *Geophysics,* 5(4):348–366, October 1940.

[103] R. Riesenfeld. On Chaikin's algorithm. *Computer Graphics and Image Processing,* 4(3):304–310, September 1975.

[104] David Salesin and Ronen Barzel. Two-bit graphics. *IEEE Computer Graphics and Applications,* 6(6):36–42, June 1986.

[105] Michael P. Salisbury, Sean E. Anderson, Ronen Barzel, and David H. Salesin. Interactive pen and ink illustration. In *Proceedings of SIGGRAPH '94,* pages 101–108. ACM, New York, 1994.

[106] Philip J. Schneider. Phoenix: An interactive curve design system based on the automatic fitting of hand-sketched curves. Master's thesis, Department of Computer Science and Engineering, University of Washington, 1988.

[107] Peter Schröder, Steven J. Gortler, Michael F. Cohen, and Pat Hanrahan. Wavelet projections for radiosity. *Computer Graphics Forum,* 13(2):141–151, June 1994.

[108] Peter Schröder and Pat Hanrahan. Wavelet methods for radiance computations. In G. Sakas, P. Shirley, and S. Müller, editors, *Photorealistic Rendering Techniques,* pages 310–326. Springer-Verlag, Berlin, 1995.

[109] Peter Schröder and Wim Sweldens. Spherical wavelets: Efficiently representing functions on the sphere. In *Proceedings of SIGGRAPH '95,* pages 161–172. ACM, New York, 1995.

[110] Thomas W. Sederberg, Peisheng Gao, Guojin Wang, and Hong Mu. 2D shape blending: An intrinsic solution to the vertex path problem. In *Proceedings of SIGGRAPH '93,* pages 15–18. ACM, New York, 1993.

[111] Thomas W. Sederberg and Eugene Greenwood. A physically based approach to 2D shape blending. In *Proceedings of SIGGRAPH '92,* pages 25–34. ACM, New York, 1992.

[112] A. Sengupta, M. Hilton, and B. Jawerth. A computationally fast wavelet-based video coding scheme. In *Digital Video Compression on Personal Computers: Algorithms and Technologies,* volume 2187 of *Proceedings of the SPIE,* pages 152–157. SPIE, Bellingham, WA, 1994.

[113] Alexander Shustorovich. Scale specific and robust edge/line encoding with linear combinations of Gabor wavelets. *Pattern Recognition,* 27(5):713–725, May 1994.

[114] François X. Sillion and Claude Puech. *Radiosity and Global Illumination.* Morgan Kaufmann, San Francisco, 1994.

[115] E. P. Simoncelli, W. T. Freeman, E. H. Adelson, and D. J. Heeger. Shiftable multi-scale transforms. *IEEE Transactions on Information Theory,* 38(2):587–607, March 1992.

[116] Mark J. T. Smith and Thomas P. Barnwell III. Exact reconstruction techniques for tree-structured subband coders. *IEEE Transactions on Acoustics, Speech, and Signal Processing,* 34(3):434–441, June 1986.

[117] Brian E. Smits. Efficient hierarchical radiosity in complex environments. Ph.D. thesis, Cornell University, 1994.

[118] Brian Smits, James Arvo, and Donald Greenberg. A clustering algorithm for radiosity in complex environments. In *Proceedings of SIGGRAPH '94,* pages 435–442. ACM, New York, 1994.

[119] Brian E. Smits, James R. Arvo, and David H. Salesin. An importance-driven radiosity algorithm. In *Proceedings of SIGGRAPH '92,* pages 85–90. ACM, New York, 1992.

[120] Eric J. Stollnitz, Tony D. DeRose, and David H. Salesin. Wavelets for computer graphics: A primer. *IEEE Computer Graphics and Applications,* 15(3):76–84, May 1995 (part 1) and 15(4):75–85, July 1995 (part 2).

[121] Michael J. Swain. Interactive indexing into image databases. In *Storage and Retrieval for Image and Video Databases,* volume 1908 of *Proceedings of the SPIE,* pages 95–103. SPIE, Bellingham, WA, 1993.

[122] Wim Sweldens. The lifting scheme: A custom-design construction of biorthogonal wavelets. Industrial Mathematics Initiative 1994:7, University of South Carolina, 1994.

[123] Wim Sweldens. The lifting scheme: A new philosophy in biorthogonal wavelet constructions. In *Wavelet Applications in Signal and Image Processing III,* volume 2569 of *Proceedings of the SPIE,* pages 68–79. SPIE, Bellingham, WA, 1995.

[124] Richard Szeliski. Fast surface interpolation using hierarchical basis functions. *IEEE Transactions on Pattern Analysis and Machine Intelligence,* 12(6):513–528, June 1990.

[125] Greg Turk and Marc Levoy. Zippered polygon meshes from range images. In *Proceedings of SIGGRAPH '94,* pages 311–318. ACM, New York, 1994.

[126] Joe Warren. Personal communication, 1994.

[127] Joe Warren. Binary subdivision schemes for functions over irregular knot sequences. In M. Daehlen, T. Lyche, and L. Schumaker, editors, *Mathematical Methods in Computer Aided Geometric Design III.* Academic Press, San Diego, 1995.

[128] Karl Weierstrass. *Mathematische Werke,* volume II. Mayer & Muller, Berlin, 1895.

[129] William Welch. Serious putty: Topological design for variational curves and surfaces. Ph.D. thesis, Carnegie-Mellon University, 1996.

[130] William Welch and Andrew Witkin. Variational surface modeling. In *Proceedings of SIGGRAPH '92,* pages 157–166. ACM, New York, 1992.

[131] Ruediger Westermann. A multiresolution framework for volume rendering. In *Proceedings of the ACM Workshop on Volume Visualization,* pages 51–58. ACM, New York, 1994.

[132] Mladen Victor Wickerhauser. High-resolution still picture compression. *Digital Signal Processing,* 2(4):204–226, October 1992.

[133] Mladen Victor Wickerhauser. *Adapted Wavelet Analysis from Theory to Software.* A. K. Peters, Wellesley, MA, 1994.

[134] Andrew Witkin and Michael Kass. Spacetime constraints. In *Proceedings of SIGGRAPH '88,* pages 159–168. ACM, New York, 1988.

[135] Andrew Witkin and Zoran Popović. Motion warping. In *Proceedings of SIGGRAPH '95,* pages 105–108. ACM, New York, 1995.

[136] Y. Q. Zhang and S. Zafar. Motion-compensated wavelet transform coding for color video compression. *IEEE Transactions on Circuits and Systems for Video Technology,* 2(3):285–296, September 1992.

INDEX

Italic page numbers indicate primary term definitions.

2-disc wavelets, 157, 163

adaptability, 3
adaptive variational modeling, 179
affine-invariance, *146*
algorithms
  binary search, 29
  Chaikin's, *62,* 63, 64, 65, 70, 72
  Christiansen-Sederberg, *128*–129
  clustering, 192–193
  Ganapathy-Dennehy, *128*–129
  image editing, 36–39
  multiresolution image querying, 44
  multiresolution tiling, *125,* 129–137
  nonstandard decomposition, 23
  optimizing tiling, *126*–128
  pyramid, 4
  wavelet radiosity, 188
analysis, *83*
  equation summary, 107
  filters, *83,* 95, 143
  k-disc, 156, 158
  lazy wavelet, 154
  matrices, 101
  single-knot, 105
  See also synthesis
approximating subdivision schemes, 65
aspect ratio, image querying, 56
averaging masks, *62,* 148–149
averaging step, *63,* 74, 144, 145
  in Loop's scheme, 146
  nonuniform subdivision, 66
  uniform subdivision, 63
  weights in, 146
  See also splitting step; subdivision

"backbone" curves, 109
base case, *130*
base mesh, *162*
  illustrated, 162
  subdividing, 162
basis functions, *13,* 16
  biorthogonal, 98
  B-spline, 92, 177–178
  Daubechies, 89
  dual, *98*–102
  Haar, *3,* 11–16
  links between, 190
  for nested spaces, 73
  normalized, *16*–18
  orthogonality, *16,* 18, 206

basis functions (*continued*)
   orthogonal multiresolution, *85*
   orthogonal wavelet, *86*
   orthonormal, *18,* 86, 98, 206
   primal, *98*
   receiving, 185
   sending, 185, 189
Bézier curves, 122
bidirectional reflectance distribution function, *193*
binary search algorithm, 29
bin function, 53
biorthogonal basis, 98
biorthogonality, 97–98
   duals and, 98–99
biorthogonal wavelets, *97–107*
   conditions defining, 108
   designing, 106–107
   single-knot wavelets, 102–105
   surface, 152–158
   See also orthogonal wavelets; semiorthogonal
      wavelets
boundary conditions, 39–40
box functions, *13*
   compactly supported, 13
   piecewise-constant, 186
   refinement, 74
   scaling, 210
box splines, 196
B-spline basis functions, 92, 177–178
   in interactive design tool, 178
   slow convergence of, 178
B-spline filter bank, 95–96
B-spline scaling functions, 91–93
   endpoint-interpolating, 92
   inner products of, 219
B-spline wavelet matrices, 209–216
   endpoint-interpolating cubic B-spline
      wavelets, 214–216
   endpoint-interpolating linear B-spline
      wavelets, 211–212
   endpoint-interpolating quadratic B-spline
      wavelets, 212–214
   Haar wavelet, 210–211
B-spline wavelets, 93–95
   endpoint-interpolating, 96
   small support of, 221

butterfly scheme, *148–149,* 163
   averaging mask for, 149
   steps of, 148
   See also surfaces

Catmull-Clark subdivision, 144
Chaikin's algorithm, *62, 63*
   for closed parametric curve, 65
   dominant left eigenvector, 72
   for a function, 63
   local subdivision algorithm for, 70
   producing uniform B-splines, 64
   See also subdivision
Christiansen-Sederberg algorithm, *128–129*
   illustrated, 129
   See also tiling problem
clustering algorithms, 192–193
coefficients
   decomposition, 47
   detail, *10*
   matching, 49
   mismatching, 49
   radiosity transport, *185,* 189
   refinement, *76*
   scaling function, 83
   splitting, 83
   wavelet, 48, 83, 142, 163
color
   correction, 39
   function, 164
   images, 30–31
color space
   CIE LUV, 56
   HSV, 47
   image querying metric, 47
   RGB, 30, 47
   TekHVC, 56
   YIQ, 30, 47, 53
compact support, *13*
compression
   curve, 120–122
   encoding, *19,* 28
   image, 28–30, 34
   polyhedral, 163
   surface, 163–164
   texture map, 164

video, 198
 wavelet, 18–20
Compress procedure, 29
conjugate gradient method, 177
content-based querying, 44, 45
control points, *64*
 Bézier, 122
 fractional-level curve, 113–114
 high resolution, 114–116
control polygon, *64*
control polyhedron, *144*
cubic B-splines, 64
 matrices, 214–216
 scaling functions, 215
 wavelets, 215
curves
 approximation, 110
 "backbone," 109
 Bézier, 122
 with changing parameterization, 118
 character of, 110, 112–113, 119
 compression of, 120–122
 editing, 109, 112–120
 fractional-level, 112
 further reading, 198
 limit, 68
 multiresolution, 109–123
 parametric in two dimensions, 119
 quartic, 177
 representations, 110–111
 role of, 109
 scan-converted, 117
 smoothing, 109, 111–112
 subdivision, 61–77
 sweep of, 109, 112–119
 See also surfaces

data structures
 multiresolution image, 34–36
 quadtree, 187
 wavelet radiosity, 187
Daubechies basis functions, 89
Daubechies subdivision, *65, 66*
Daubechies wavelets, 88–89
Decompose function, *37–38*
Decomposition procedure, 17

decompositions, *37*
 coefficients, 47
 computing, 47
 edge information and, 47
 image approximation and, 47
 nonstandard, *23–25*, 47
 of polyhedral surface, 143
 standard, *21–23*, 47
 transport matrix, 190–191
 type decision, 47
DecompositionStep procedure, 17
derivative masks, *69*
detail coefficients, *10*
differencing, 15
dilation, *75*
direct manipulation, *116*
 editing with, 116–117
 multiresolution editing and, 118
discrete radiosity transport equation, *184–185*
Display routine, 36–37, 40
DLG scheme, *66, 67*
dominant left eigenvector, *71–72*
 Chaikin's algorithm, 72
 limit position, 72
 See also evaluation masks
Doo-Sabin subdivision, 143, 144
DrawSegment procedure, 121
dual basis functions, *98*, 184
 biorthogonality and, 98–99
 constructing, 102
 defining through subdivision, 99–102
 unnormalized, 100
 See also biorthogonal wavelets
dual lifting, *106*
dual scaling functions, 100, 106
dual subdivision matrices, 101
dyadic points, 62

edge swaps, *134–135*
editing
 with direct manipulation, 116–117
 image, 33–41
 image compression and, 34
editing curves, 109, 112–120
 character, 119
 fractional-level, 113–116

editing curves (*continued*)
  multiresolution, 118, 165–166
  orientation of detail, 119–120
  sweep, 112–119
  See also curves
eigenstructure, *70*
eigenvalues, *207*
eigenvectors, *207–208*
encoding, *19,* 28
endpoint-interpolating B-splines, 67, 68
  cubic, 67, 94
  cubic, matrices, 214–216
  knot insertion, 67
  linear, matrices, 211–212
  nonuniform, 91
  quadratic, matrices, 212–214
  scaling functions, 75
  wavelets, 96
ErrBound function, 121–122
evaluation masks, *69–72*
  averaging, 69
  derivative, 69
  dominant left eigenvector, *71–72*
  local subdivision matrix, 70–71
  for surfaces, 147
  See also subdivision
Extrapolate procedure, 38–39
extrapolation, *37,* 38–39
  compositing operations and, 38
  routine, 38–39

face schemes, *144*
  quadrilateral, *144–145*
  triangular, *144*
  See also splitting step
father scaling function, *75*
  refinement relation for, 76
  See also scaling functions
filter bank, *11,* 82–85
  B-spline, 95–96
  illustrated, 84
  sequence coefficients, 130
filter-bank reconstruction, *130–132*
  illustrated, 132
  steps for, 132

suspect edge list, 134
  See also multiresolution tiling algorithm
final gathering step, *191–192*
finite-dimensional spaces, *204*
finite-element method, *173*
finite elements, *174*
  physical simulation problems and, 199
  radiosity and, 183–186
  in variational modeling, 173–177
flatlets, 186
Fourier analysis, 1–2
fractional-level curves, 112
  control points, 113–114
  editing, 113–116
  See also curves
fractional resolutions, 40
fractional zooming, 40
functions
  bidirectional reflectance distribution, *193*
  bin, 53
  box, *13,* 74, 186
  ErrBound, 121–122
  hat, 102
  importance, 192
  interpolation, 115
  limit, 73, 75
  objective, *171–173*
  one-dimensional, 11–16
  piecewise-constant, 11–12
  quartic, 174, 175
  scaling, *13,* 25, 73
  See also basis functions
further reading, 195–200
  curve applications, 198
  image applications, 197–198
  multiresolution analysis theory, 195–196
  physical simulation, 199–200
  surface applications, 198

Ganapathy-Dennehy algorithm, *128–129*
  illustrated, 129
  See also tiling problem
Gauss-Seidel method, 177
global illumination, *181–194*
  enhancements to wavelet radiosity, 192–194

further reading, 199
glossy, 193
radiosity problem, *181*–186
view independent, 192
wavelet radiosity, 186–192
See also physical simulations; radiosity
glossy global illumination, 193

Haar basis functions, *3,* 16
  normalization, *16*–18
  one-dimensional, 11–16
  orthogonality, *16,* 18
  two-dimensional, 25–27
Haar wavelets, *14*
  B-spline matrices, 210–211
  illustrated, 15
  image editing and, 33–41
  one-dimensional transforms, 9–11
  two-dimensional transforms, 21–25
half-open intervals, *11*
hat functions, 102
Hessian matrix, *175*–*176*
hierarchical B-splines, 110, *141*
hierarchical radiosity, *186,* 192
hierarchical representations. *See* multiresolution
    methods
homogeneous system, *91*

identity mask, 64
image approximation, 47
image clusters, 56
image compression, *28*–30
  editing and, 34
  illustrated, 30
  step summary, 28
  See also wavelet compression
image databases, 34
  access to, 43
  indexing with keywords, 43
image editing, 33–41
  algorithm, 36–39
  boundary conditions and, 39–40
  examples, 41
  at fractional resolutions, 40
  on compressed images, 34
image painting and composition system, 33

image querying, 43–57
  algorithm, 50–53
  aspect ratio, 56
  average times, 56
  by content, 45
  content-based, 44, 45
  examples, 53–55
  extensions, 55–57
  image clusters, 56
  "interactive mode," 54–55
  keys, 56
  multiresolution algorithm, 44, 45
  partial, 57
  perceptually based spaces, 56
  preprocessing step, 50–51
  querying step, 51–53
  target image signature comparison, 52
  video, 57
image querying metric, *44,* 48–49
  color space, 47
  components of, 47–48
  "correct," 46
  decomposition type, 47
  development, 46–50
  effectiveness, 54
  fast computation of, 49–50
  final, 49
  illustrated, 49
  multiple, 57
  multiresolution approach, 46–47
  normalization, 48
  quantization, 48
  truncation, 48
  wavelet type, 47
  weights, 54
  See also image editing
images, 11
  color, 30–31
  multiresolution, *34*–36
  one-dimensional, 12
  query, 44
  target, 44, 46
  two-pixel, 11
importance functions, 192
"infinite desktop" user-interface metaphor, 34
infinite-dimensional spaces, *204*

inner product, *13–14,* 93
    of Bernstein polynomials, 219
    of B-spline scaling functions, 219
    precomputation of, 151
    review, 205
    selecting, 96, 151
    surface wavelet, 151
inner product space, *205*
interactive paint systems, 34
interpolating schemes, 65–66
invertibility
    equation summary, 107
    identities, 115

k-disc wavelets, *156*
    analysis matrices, 156, 158
    synthesis matrices, 156, 158
knot insertion, *217*

Lagrange multipliers, *176*
lazy wavelets, *102–104*
    analysis, 154
    illustrated, 103
    improving, 104
    practical use and, 104
    problems with, 152
    scaling functions, 152
    synthesis matrices, 153, 154
left eigenvectors, *71–72, 207*
level-of-detail control, *164*–165
lifting operation, 98, *106*
    biorthogonal bases and, 98
    dual, *106*
limit curves, 68
limit function, 73, 75
linear algebra, 203–208
    bases and dimension, 204
    eigenvectors and eigenvalues, 207–208
    inner products and orthogonality, 205–206
    norms and normalization, 206–207
    vector spaces, 203–204
linear B-splines, 64
linear interpolation, 118
linear-time complexity, 3
Link structure, 187
local optimization, 134–135

local subdivision matrix, *70–71*
    for Chaikin's algorithm, 70
    eigenanalysis of, 70
Loop's scheme, *145–147*
    averaging step in, 146
    masks, 147
    See also surfaces

masks, *62,* 64–66, *146–149*
    butterfly scheme, 149
    Chakin's algorithm, 62
    Daubechies subdivision, 65
    DLG scheme, 66
    Loop's scheme, 147
    tangent, 147
    uniform B-splines, 64
    See also evaluation masks
matching coefficients, 49
Matlab code, 209, 217
mismatching coefficients, 49
mother wavelet, *81*
multiresolution analysis, *4,* 80–85
    in arbitrary topological domains, 142
    for B-splines, 111
    filter bank, 82–85
    framework, 111
    further reading, 195–196
    refinement, 80–82
    shift-invariant, 4
    shift-variant, 4–5, 79, 196
    starting point, 79, 80
    for surfaces, 142–143
    theory of, 79–108
    wavelet space definition, 80
multiresolution curves, 109–123
    related representations, 110–111
    representation, 110
    See also curves
multiresolution editing, 165–166
    direct manipulation and, 118
    examples, 166
    surface application, 166
    See also editing curves
multiresolution image querying algorithm, 44
multiresolution images, *34*
    applications, 34

data structures, 34–36
  storage, 34
multiresolution methods, *2–3*
multiresolution painting, 36–39
  display, 36–37
  painting, 37
  update, 37–39
multiresolution tiling algorithm, *125,* 129–137
  base case tiling optimization, 130
  contour decomposition, 130
  edge swaps, *134*–135
  filter-bank reconstruction, *130*–132
  illustrated steps, 131
  key steps in, 129
  local optimization, 134–135
  reconstruction, 130–134
  single-wavelet reconstruction, *131,* 132–134
  tiling examples, 136–137
  tiling produced by, 130
  time complexity, 135–136
  See also tiling problem
multiwavelets, 186

nested spaces
  basis functions for, 73
  chain of, 151
  refinable scaling functions and, 72–77,
    149–151
nondefective matrix, *207*
nonstandard construction, *25, 27*
  advantages, 27
  illustrated, 27
  square supports, 27
  See also standard construction
nonstandard decomposition, *23–25,* 47
  advantages, 27
  algorithm, 23
  illustrated, 24
  reconstruction algorithm, 24–25
  See also standard decomposition
NonstandardDecomposition procedure, 23
NonstandardReconstruction procedure, 25
nonuniform subdivision, *66–68*
  averaging step, 66
  See also subdivision
norm, *206*

normalization, *16*–18, 206
  image querying metric, 48
  pseudocode procedures, 17–18
normalized tangents, 120
null space, *91*

objective function, *171*
  defining, 172
  setting up, 172–173
  See also variational modeling
one-dimensional Haar basis functions, 11–16
one-dimensional Haar wavelet transform, 9–11
one-dimensional images, 12
optimizing algorithm, *126–128*
  drawback to, 127–128
  illustrated results, 128
  illustrated tiling, 127
  user interaction and, 128
  See also tiling problem
oracle, *179,* 189
orthogonality, *16*
  implications of, 86–88
orthogonal matrix, *87*
orthogonal multiresolution basis, *85*
orthogonal wavelet basis, *86*
orthogonal wavelets, *85–89*
  conditions defining, 108
  Daubechies wavelets, 88–89
  orthogonality implications, 86–88
  synthesis matrix, 87
  See also biorthogonal wavelets; semiorthogo-
    nal wavelets; wavelets
orthonormal basis, *18,* 85, 86, 98, 206

painting, 37
  binary operations, 38
  multiresolution implementation, 36–39
pairwise averaging, 15
photorealistic image synthesis, 181
physical simulations, 34
  further reading, 199–200
  global illumination, 181–194
  variational modeling, 171–179
piecewise-constant functions, 11–12
piecewise-linear scaling functions, 103
polyhedral compression, 163

polyhedral subdivision, *145,* 148
    analysis matrices for, 154–155
    of tetrahedron, 153
    See also subdivision
polyhedral surface wavelets, 156, 157
"popping," 165
Powers of Ten, 41
preprocessing step, 50–51
pre-wavelets, *14,* 90
    See also semiorthogonal wavelets
primal basis, 98
progressive transmission, *165*
pseudo-inverse matrices, 116
pyramid algorithms, 4

QBIC, *45*
quadratic programming, *176*
quadrature mirror filters, 4, *88*
quadtree, 34–*36*
    data structure, 35–36, 187
    node information, 35
    node transparency, 35
    sparse, 36, 38
QuadTreeNode structure, 35, 187
QuadTree structure, 36, 187
quantization, wavelet coefficient, 48
quartic curve, 177
quartic functions, 174, 175
query by visual example (QVE), *45*
query images, 44
querying. *See* image querying

radiance, *193*
    generalization to, 193–194
    transport equation, *193*
    See also global illumination
radiosity, *181*–194
    discretizing, 184
    distribution, 184
    emitted, *182*
    finite elements and, 183–186
    hierarchical, *186,* 192
    wavelet, 186–192
    See also global illumination
radiosity transport
    coefficients, *185,* 189

discretizing, 184–185
    equation, *182*
    geometry, 183
    matrix, computing, 185
    matrix decompositions, 190–191
    operator, *182*
reconstruction, 130–134
    filter-bank, 130–132
    of original tilings, 135
    single-wavelet, *131,* 132–134
    See also multiresolution tiling algorithm
Reconstruction procedure, 17
ReconstructionStep procedure, 18
recursive subdivision. *See* subdivision
refinable scaling functions, *73*
    nested spaces and, 72–77, 149–151
    See also scaling functions; subdivision
refinement, *75*
    coefficients, *76*
    equation summary, 107
    for father scaling function, 76
    goal of, 188
    steps, 144
    wavelet radiosity, 188–190
RGB colors, 30, 47
right eigenvectors, *207*

scaling functions, *13, 73,* 81
    box, 210
    B-spline, 91–93, 219
    coefficients, 83
    computing, 75
    cubic B-spline, 215
    dual, 100, 106
    father, *75*
    lazy wavelet, 152
    linear B-spline, 211
    linear combinations of, 72
    piecewise-constant B-spline, 210
    piecewise-linear, 103
    quadratic B-spline, 213
    refinable, *73*
    refinement coefficients for, *76*
    refinement matrix for, 77
    refinement relation for, *75,* 75

synthesis matrix, 217
two-dimensional, 25
two-scale relation for, *82*
uniform subdivision, 76
See also subdivision
scaling, levels of, 26
scan conversion, 120–123
bound, 121
curve compression and, 120–122
within error tolerance, 123
See also curves
ScoreQuery function, 52–53
second-generation wavelets, *196*
self-dual, *184*
semiorthogonal wavelets, *90–97*
conditions defining, 108
designing, 96–97
implications, 90–91
spline wavelets, *91–96*
See also biorthogonal wavelets; orthogonal
wavelets
shiftable transforms, *197*
shift-invariant multiresolution analysis, *4*
shift variant multiresolution analysis, *4–5, 79*
further reading, 196
See also multiresolution analysis
signatures, *44–45*
single-knot wavelets, *104–105*
analysis/synthesis sequences, 104–105
construction of, 105, 106
synthesis matrix, 105
single-wavelet reconstruction, *131, 132–134*
illustrated, 134
steps for, 133
suspect edge list, 134
See also multiresolution tiling algorithm
smoothing curves, 109, 111–112
continuously, 112
See also curves
spacetime constraints, *199–200*
sparsity, 3
spline wavelets, *91–96*
B-spline filter bank, 95–96
B-spline scaling functions, 91–93
B-spline wavelets, 93–95
inner product, 93

splitting step, *63, 74*
for triangular subdivision, 145, 146
types of, 144
See also averaging step
standard construction, *25, 26–27*
advantages, 27
contents of, 25
illustrated, 26
square supports, 27
See also nonstandard construction
standard decomposition, *21–23, 47*
advantages, 27
algorithm, 22–23
illustrated, 22
reconstruction algorithm, 23
See also nonstandard decomposition
StandardDecomposition procedure, 22–23
StandardReconstruction procedure, 23
steerable filters, *197*
steradian, *193*
subdivision, *61*
approximating schemes, *65*
averaging step, *63*
base mesh, 162
butterfly, 148–149
Catmull-Clark, 144
for characterizing functions, 100
Daubechies, *65,* 66
defining duals through, 99–102
DLG scheme, *66,* 67
Doo-Sabin, 144, 145
implementing, 62
interpolating schemes, 65–66
linear B-spline, 65
local matrix, *70–71*
Loop's scheme, 145–148
matrix, *74,* 77
nonuniform, *66–68, 196*
polyhedral, *145,* 148, 153
shift-variant setting based on, 196
splitting step, *63*
stationary, 100, 196
surfaces, 143–151
triangular, 145
uniform, *62–66*
uses, 61

subdivision curves, 61–77
    evaluation masks, 68–72
    refinable scaling functions, 72–77
support, *13*, 27, 79, 89, 94, 100, 155–156, 185,
    221
surface compression, 163–164
    polyhedral, 163
    texture map, 164
surfaces
    applications, 161–167
    approximations, 163
    conversion to multiresolution form, 161–162
    evaluation masks for, 147
    further reading, 198
    multiresolution analysis for, 142–143
    multiresolution representations of, 158–159
    shaded images of, 148
    subdivision, 143–151
    See also curves
surface wavelets, 141–159
    biorthogonal construction, 152–158
    future directions for, 166–167
    inner product selection, 151
    polyhedral, 156, 157
    surface subdivision and, 143–151
synthesis, *83*
    equation summary, 107
    filters, *83,* 97, 143
    k-disc, 156, 158
    single-knot, 105
    wavelet matrices, 97, 101
    See also analysis

tangent masks, *147*–148
    See also evaluation masks; masks
target images, 44, 46
    comparing to query signature, 52
    See also image querying
tensor-product construction, 142
texture generation, 197
texture map compression, 164
texture mapping, 34
tiling problem, *125*–129
    Christiansen-Sederberg algorithm, 128–129
    difficulty of, 125

Ganapathy-Dennehy algorithm, *128*–129
    optimizing algorithm, *126*–128
    previous solutions to, 126–129
    See also multiresolution tiling algorithm
truncation, wavelet coefficient, 48
two-dimensional Haar basis functions, 25–27
    nonstandard construction, *25,* 27
    standard construction, *25,* 26–27
    See also Haar basis functions
two-dimensional Haar wavelet transforms, 21–25
    nonstandard decomposition, *23*–25
    standard decomposition, *21*–23
two-scale relation, *82*

uniform subdivision, *62*–66
    refinement relation, 76
    scaling functions, 76
    See also subdivision
unnormalized tangents, 120
update operation, 37–39
    decomposition, 37–38
    extrapolation, 37–39

vanishing moments, *90*
variational modeling, *171*–179
    adaptive, 179
    finite elements in, 173–177
    further reading, 199
    using wavelets, 177–179
    See also objective function; physical simula-
        tions
vectors, 11–14, *204*
    linearly independent, *204*
    normalized, *206*
    orthogonal, 14
    types of, 204
vector spaces, *11*–13, 203–204
    basis for, *13*
    finite-dimensional, *204*
    infinite-dimensional, *204*
    inner product, *13–14*
    nested, 13
    orthogonal complement, 14
    review of, 203–204
vertex schemes, *144*
    See also splitting step

video compression, 198
video querying, 57
virtual reality, 34

wavelet basis
  compact support, 89
  orthogonal, *86*
  for radiosity, 186
  smoothness, 90
  symmetry, 90
  vanishing moments, 90
wavelet coefficients, 48, 83, 142, 163
wavelet compression, 18–20
  applying, 20
  goal of, 18
  problem, 19
  varying amounts of, 20
  See also image compression
wavelet decomposition. *See* wavelet transform
wavelet packet transform, *197*
wavelet radiosity, 186–192
  algorithm, 188
  clustering, 192–193
  data structures, 187
  enhancements to, 192–194
  final gathering step, *191*–192
  generalization to radiance, 193–194
  importance-driven refinement, 192
  refinement, 188–190
  wavelet basis for, 186
  See also radiosity
wavelets, 2, *14,* 80
  2-disc, 157, 163
  adaptability, 3
  applications of, 1, 2
  biorthogonal, *97*–107
  B-spline, 93–95, 96, 211, 213, 215

cubic B-spline, 215
Daubechies, 88–89
Haar, *14*
historical perspective, 3–5
k-disc, *156*
lazy, *102*–104
linear B-spline, 211
linear-time complexity, 3
orthogonal, *85*–89
polyhedral surface, 156, 157
quadratic B-spline, 213
second-generation, *196*
semiorthogonal, *90*–97
single-knot, *104*–105
sparsity, 3
spline, *91*–96
surface, 141–159
vanishing moments, *90*
wavelet spaces
  defining, 80
  wavelet bases for, 91
wavelet transform, *10,* 84
  advantages of, 11
  color information and, 30
  computing, 11
  Haar, 16
  one-dimensional, 9–11
  storing, 11
  two-dimensional, 21–25
weights
  in averaging step, 146
  image querying metric, 54

YIQ color space, 30, 47, 53

zooming, fractional, 40

Related Titles from Morgan Kaufmann:

Principles of Digital Image Synthesis

by Andrew S. Glassner, Microsoft Research

A comprehensive presentation of the three core fields of study that constitute digital image synthesis: the human visual system, digital signal processing, and the interaction of matter and light. Assuming no more than a basic background in calculus, Glassner demonstrates how these disciplines are elegantly orchestrated into modern rendering techniques such as radiosity and ray tracing.

1995; 1600 pages; two-volume set; cloth; ISBN 1-55860-276-3

Interactive Curves and Surfaces

by Alyn Rockwood, Arizona State University and Peter Chambers, VLSI Technology, Inc.

An innovative and intuitive computer-aided geometric design (CAGD) laboratory that runs under Windows. This book/disk package provides an understanding of the standard ways of creating curved lines and surfaces, the mathematical tools behind the generation of curves and surfaces, and the development of computer-based CAGD algorithms. It also provides powerful interactive test benches to explore the behavior and characteristics of the most popular CAGD curves.

1996; book/disk package; 300 pages; paper/3.5" disk; ISBN 1-55860-405-7

Jim Blinn's Corner: A Trip Down the Graphics Pipeline

by Jim Blinn, Microsoft Research

A compendium of the best columns from "Jim Blinn's Corner," graphic guru Blinn's regular column in *IEEE Computer Graphics and Applications*. With its added commentary, this book offers special tips and tricks for programmers, with emphasis on the mathematical aspects of computer graphics.

1996; 300 pages; paper; ISBN 1-55860-387-5